The Non-Fundamentality
of Spacetime

This book argues that our current best theories of fundamental physics are best interpreted as positing spacetime as non-fundamental. It is written in accessible language and largely avoids mathematical technicalities by instead focusing on the key metaphysical and foundational lessons for the fundamentality of spacetime.

According to orthodoxy, spacetime and spatiotemporal properties are regarded as fundamental structures of our world. Spacetime fundamentalism, however, faces challenges from speculative theories of quantum gravity – roughly speaking, the project of applying the lessons of quantum mechanics to gravitation and spacetime. This book demonstrates that the non-fundamentality of spacetime does not rely on speculative physics alone. Rather, one can give an interpretation of general relativity that supports some form of spacetime non-fundamentalism. The author makes the case for spacetime non-fundamentalism in three steps. First, he confronts the standard geometrical interpretation of general relativity with Brown and Pooley's dynamical approach to relativity theory. Second, he considers an alternative derivation of the Einstein field equations, namely the classical spin-2 approach, and argues that it paves the way for a refined dynamical approach to general relativity. Finally, he argues that particle physics can serve as a continuity condition for the metaphysics of spacetime.

The Non-Fundamentality of Spacetime will be of interest to scholars and advanced students working in philosophy of physics, philosophy of science, and metaphysics.

Kian Salimkhani is a Postdoctoral Researcher at the University of Cologne, Germany, and a member of the DFG research group "Inductive Metaphysics". He mainly works on philosophy of physics, metaphysics, and general philosophy of science.

Routledge Studies in the Philosophy of Mathematics and Physics

Edited by Elaine Landry, University of California, Davis, USA and Dean Rickles, University of Sydney, Australia

For more information about this series, please visit: https://www.routledge.com/
Routledge-Studies-in-the-Philosophy-of-Mathematics-and-Physics/book-series/PMP

The Non-Fundamentality of Spacetime

General Relativity, Quantum Gravity, and Metaphysics

Kian Salimkhani

Routledge
Taylor & Francis Group

NEW YORK AND LONDON

First published 2024
by Routledge
605 Third Avenue, New York, NY 10158

and by Routledge
4 Park Square, Milton Park, Abingdon, Oxon, OX14 4RN

Routledge is an imprint of the Taylor & Francis Group, an informa business

© 2024 Kian Salimkhani

ISBN: 978-1-032-51833-6 (hbk)
ISBN: 978-1-032-51834-3 (pbk)
ISBN: 978-1-003-40414-9 (ebk)

DOI: 10.4324/9781003404149

Typeset in Sabon
by KnowledgeWorks Global Ltd.

To Mona

Contents

Acknowledgements

This book is based on my dissertation defended at the University of Bonn on 22 December 2020.

First and foremost, I would like to thank my supervisor Andreas Bartels, who continuously provided excellent support, encouraging advice, and the freedom to independently find and pursue my own research projects. His genuinely interested and detailed feedback has, without a doubt, contributed greatly to improving my work and its presentation. I also thank the other members of my examination committee – Elke Brendel, Andreas Hüttemann, and Dennis Lehmkuhl.

A big thank you to Nick Huggett for his kind support during preparing the book proposal and to an anonymous reviewer with Routledge for reading and recommending it for publication. I also wish to give special thanks to Andrew Weckenmann and Rosaleah Stammler from Routledge for their very supportive and incredibly fast and reliable work as well as Dean Rickles and Elaine Landry for kindly agreeing to include this book in their *Routledge Studies in the Philosophy of Mathematics and Physics*. Many thanks also to the production team.

For their efforts to listen to, read, and discuss some of the passages of this book that have been previously presented and published, I am very thankful to many friends and colleagues. I especially thank Andreas Bartels, Kaća Bradonjić, Maren Bräutigam, Harvey Brown, Karen Crowther, Radin Dardashti, Olivier Darrigol, Richard Dawid, Juliusz Doboszewski, Cord Friebe, Stefan Heidl, Vera Hoffmann-Kolss, Nick Huggett, Andreas Hüttemann, Rasmus Jaksland, Lucy James, Vincent Lam, Dennis Lehmkuhl, Niels Linnemann, Niels Martens, Vera Matarese, Keizo Matsubara, Damián Kaloni Mayorga Peña, Kerry McKenzie, Tushar Menon, Paul Oehlmann, Brian Pitts, Oliver Pooley, Carina Prunkl, James Read, Sébastien Rivat, Christian Röken, Matthias Rolffs, Thorsten Schimannek, Manfred Stöckler, Martin Voggenauer, Isaac Wilhelm, Alastair Wilson, Christian Wüthrich, and Ben Young. To many others I am grateful for helpful discussions as well and for time spent together at various conferences, workshops, and summer schools.

I also thank my students at the Universities of Bonn and Cologne, and the participants of the Weekend Seminars on Philosophy of Physics, which I have the pleasure to co-organise. Through teaching them, I learned a lot myself. For the numerous, truly memorable seminars during my own studies that have renewed my passion for philosophy, I thank Guido Kreis.

Thank you to the (former) members of the Institute for Philosophy in Bonn; special thanks go to all former members of Andreas Bartels' Chair, especially Stefan Heidl and Matthias Rolffs for memorable times in our joint office. I also would like to express my gratitude to the members of the DFG research group *Inductive Metaphysics*, the Doctoral Programme of the Faculty of Arts at the University of Bonn, and Dennis Lehmkuhl's Lichtenberg group for History and Philosophy of Physics. For a nice time at the Philosophical Seminar in Cologne, I thank Andreas Hüttemann and his group, especially Thomas Blanchard, Maren Bräutigam, Christian Feldbacher-Escamilla, Farzaneh Hassanali, Anastasja Petrović, Bram Vaassen, and Martin Voggenauer. For much-needed administrative help in Bonn and Cologne, I thank Martina Richtberg and Ursula Heister. For the joint organisation of the Weekend Seminars on Philosophy of Physics and the Quarterly Lectures on Philosophy of Science, I thank Maren Bräutigam, Niels Linnemann, Anastasja Petrović, Oxana Shaya, Annica Vieser, and Karla Weingarten.

With respect to funding, I thank the Deutsche Forschungsgemeinschaft (DFG, German Research Foundation), research group *Inductive Metaphysics*, grant number FOR 2495; and the Doctoral Programme of the Faculty of Arts at the University of Bonn. In addition, I am grateful to Christian Wüthrich for a Beyond Spacetime Junior Visiting Fellowship at the University of Geneva in May 2018 and to the Centre for Philosophy and the Sciences at the University of Oslo for a CPS Visiting Scholarship in 2023.

Special thanks go to Niels Linnemann and Rasmus Jaksland for their repeated feedback on different parts of this work and, probably even more importantly, their moral support. Similarly, special thanks to Stefan Heidl and Matthias Rolffs for repeated discussions of various philosophical points and moral support.

A big thank you to all my friends, some of whom have already been named.

Finally and most importantly, I am deeply thankful to my parents, my brother, and my whole family for their unconditional support and their patience. In particular, my greatest thanks go to Mona for really just everything – thank you!

1   Introduction

We are used to speaking of entities and events in space and time (or more precisely, in spacetime). But in what way does spacetime exist, i.e., what ontological status does it have? Does it exist distinctly and independently of the material entities and events in it, virtually as a stage, as substantivalism claims? Or, does spacetime exist as dependent on these entities and events, as relationalism claims? In other words, how do the spatiotemporal and the material structures of the world relate to each other? This book seeks to answer these traditional questions, which are once again the subject of intense debate in contemporary philosophy. My thesis is that central aspects of spacetime are reducible to material structures and their properties. Thus, the book can be read as an attempt to reformulate a relational understanding of spacetime broadly construed.

Reviving the substantivalism–relationalism debate, which has already been conducted several times since Newton, Clarke, and Leibniz, is pressing for at least three reasons. First, the standard position, that is, a form of substantivalism based on general relativity, is currently facing some headwind from physics, namely from theories of quantum gravity. Many physicists and philosophers are convinced that these new theories, which, roughly speaking, try to apply the lessons of quantum mechanics to gravity and spacetime, will reduce spacetime to some non-spatiotemporal quantum structure. It is controversial, though, how the speculative character of these hitherto empirically unconfirmed theory proposals should be dealt with. Second, as I shall argue in this book, promising approaches for undermining substantivalism emerge from the recent debate about explanations in special and general relativity. Third, modern metaphysics, with its theories of fundamentality and grounding, now provides a conceptual apparatus that facilitates a new perspective on the substantivalism–relationalism debate.

As I elaborate in Chapter 2 ("Fundamentality"), it is through this new conceptual apparatus that the central question of the substantivalism–relationalism debate can be made more precise. The traditional debate

DOI: 10.4324/9781003404149-1

focuses on the question of existence; the modern debate focuses on the question of fundamentality. Asking whether some entity is fundamental is not just about asking whether some entity is on the list of what there is, but about examining its ontological status *with respect to other entities*. The fundamental entities are *ontologically privileged* over all other entities. They are *ontologically prior* to them. Thus, while traditionally it was in question whether spacetime is part of the ontology of our world at all, now the focus shifts to whether spacetime enjoys an ontological priority: for even if spacetime is part of the ontology, it remains to be answered whether it is part of the fundamental ontology, i.e., whether it belongs to the subset of entities that are ontologically prior. I first present mainstream analyses of the notion of fundamentality and conclude that, for my purposes, it is appropriate to understand fundamentality primarily as *ontological independence*. Here, independence should not get confused with isolation. Ontologically independent entities do stand in multiple ontological dependence relations to other entities. However, all these ontological dependence relations run in one direction: the ontologically independent entities determine the ontologically dependent entities, not *vice versa*.

Specifying debates in the philosophy of spacetime in terms of fundamentality has many advantages. First, the input from new physical theories is more appropriately understood and appreciated: new physical theories do not merely provide an update on what entities exist, but above all shed new light on the dependence relations between the entities. Second, shifting the attention to dependence relations guards against a formalistic or literal theory interpretation which posits fundamentals primarily because they appear in the relevant equations. For often the formalism merely articulates symmetric dependence relations that do not sanction any immediate inference to what is fundamental. Third, in this specific case, two previously largely separate debates about the interpretation of relativity theory converge in a very fruitful way if we focus on dependence relations. In general, integrating the metaphysical literature helps to clarify and order the debates in the philosophy of spacetime that are usually carried out without taking sufficient notice of said metaphysical concepts.

To account for this refinement of the debate, I shall call the standard position *spacetime fundamentalism*. In a nutshell, spacetime fundamentalism claims that spacetime is fundamental, i.e., ontologically independent.

As I explain in detail in Chapter 3 ("Spacetime in Relativity Theory"), modern spacetime fundamentalism is justified primarily on the grounds of general relativity – our best physical theory of spacetime. Very briefly, one of the main arguments can be summarised as follows: our spacetime, i.e., according to empirical data, that of general relativity, is fundamental because (1) general relativity formally introduces spacetime as an entity independent from material structures and (2) in the standard interpretation of general relativity, spacetime serves as a reduction basis for something else, namely gravity. Gravity receives the status of a pseudo-force in general relativity: gravity is reduced to the curvature of spacetime. Unlike the electromagnetic force, for example, gravity is not represented by its own physical force field. Instead, the metric field g, which together with the manifold represents general-relativistic spacetime, also accounts for gravity. General relativity *geometrises* gravitation.

The geometrical interpretation is supported by the equivalence principle, which arguably represents the empirical core of general relativity. The exact content of the equivalence principle is the subject of an ongoing debate, to which this book offers a contribution, but it can be roughly outlined as follows: it is impossible to detect the presence of a gravitational field by local experiments. Thus, the validity of the equivalence principle indicates that gravity – unlike electromagnetism, for example – is a universal force that affects all entities with mass or energy. There are no gravitationally neutral objects with respect to which we could measure local gravitational accelerations. In fact, gravity even affects all entities in the same way, regardless of their properties. Put briefly, it is this universality of the gravitational acceleration that establishes the geometrical understanding of gravitation and thus, indirectly, the fundamentality of spacetime. The fact that gravity's effect on the material entities does not depend on the properties of these entities is best explained, so the claim, by the fact that gravity stems from the ontologically independent spacetime itself. As a result, the fundamentality of spacetime would be corroborated, at least with regard to general relativity: spacetime is fundamental since this explains gravity and its universality.

Now, the plausibility of this argumentation largely owes to an assumption that is not explicitly justified itself: that the effect of spacetime on all material entities is the same. This assumption might be challenged. It becomes apparent that, ultimately, specifying the dependence relations between gravity and spacetime geometry exceeds the explanatory resources of general relativity. Determining an already established dependence relation more precisely, in particular, determining its direction, needs to draw on additional assumptions. The problem of justifying the direction of a given dependence relation also arises in the course of my argumentation. With respect to said critique, however, the spacetime fundamentalist can rightfully point out that the explanatory deficit (the unjustified assumption) does not bring down their argumentation yet. After all, it was only claimed that it is the best explanation and not that it is the perfect explanation.

Against this background, spacetime fundamentalism is mainly threatened by theories of quantum gravity – and thus by the insight that general relativity will probably not turn out to be the correct theory of spacetime in the end. However, the multitude of theory proposals with very different ontological implications and the immense problem that, for the time being, no empirical data can be expected to test these theories make this attempt to undermine spacetime fundamentalism vulnerable. In this book, I therefore try to show that the non-fundamentality of spacetime does not follow from speculative new physics alone. Established, i.e., empirically well-confirmed, theories support it as well. Amongst other things, I present an interpretation of general relativity that lends credence to a form of spacetime non-fundamentalism.

Specifically, I argue against spacetime fundamentalism in three steps. First, I show that general-relativistic spacetime is at least not entirely independent of matter properties. Second, I extend this argument using the so-called classical spin-2 theory, an empirically equivalent reformulation of general relativity (up to a few qualifications). This creates a tie between spacetime fundamentalism and spacetime non-fundamentalism. To resolve the tie in favour of spacetime non-fundamentalism, I finally invoke the overall context of physics, including quantum physics, in the third step.

For the first step, I confront the standard geometric interpretation of special and general relativity with an alternative interpretation: the dynamical approach by Harvey Brown, Oliver Pooley, and James Read. Centrally, the dynamical approach rejects the explanatory relations between spacetime geometry and matter dynamics presented by the geometrical view as unsatisfactory. I examine the dynamical approach in detail and argue that, while it does not refute spacetime fundamentalism, it provides promising avenues for its rejection by highlighting dependence relations between properties of spacetime and properties of matter. The geometrical–dynamical debate about explanations in relativity theory reveals that spatiotemporal distances can only be measured with material rods and clocks if the symmetry properties of spacetime coincide with those of matter. Assuming that both structures are ontologically independent blocks an explanation of the fact that spatiotemporal distances can indeed be measured with material rods and clocks – this would have to be considered an inexplicable coincidence.

These explanatory issues, however, do not immediately affect the ontological question of whether general-relativistic spacetime is fundamental. In particular, also Brown, Pooley, and Read take general-relativistic spacetime as a fundamental entity. The underlying idea can be reconstructed as follows: some properties of spacetime do depend on matter structures, the core entity (represented by metric and manifold), however, is very much ontologically independent. Thus, the dynamical interpretation of general relativity is in principle compatible with spacetime fundamentalism. Interestingly, this is different in the case of special relativity. Here, the special-relativistic spacetime as such (more precisely: the spacetime metric) is reduced to symmetry properties of matter dynamics. According to Brown and Pooley, the special-relativistic spacetime is therefore completely eliminated from the ontology. It is a "non-entity". For my purposes, it suffices for the time being that the special-relativistic spacetime is thus implicitly non-fundamental as well – later, I shall explicitly argue that realism is not at stake here.

Besides the ontological status of metrical aspects of spacetime (represented by the spacetime metric), I also address the ontological status of topological aspects of spacetime (represented by the manifold) in a detailed digression based on joint work with Niels Linnemann. The proponents of the dynamical approach typically do not disclose what ontological status and explanatory role they ascribe to the manifold. This gives rise to what we call the problem of pregeometry. I discuss various options for dealing with this problem and conclude that it is solved by conceiving of the manifold as a non-entity – so here, it actually is about realism.

Finally, from analysing why the ontological reduction of the spacetime metric succeeds in the special-relativistic but not in the general-relativistic case, a starting point for overcoming this problem emerges. In the second step of my argumentation against spacetime fundamentalism, I demonstrate in Chapter 4 ("Classical Spin-2 Gravity") that an alternative formalism, namely classical spin-2 theory, helps to advance the dynamical approach to general relativity in such a way that an ontological reduction of general-relativistic spacetime to matter structures is feasible. The key is that there is a sense in which the dependent and independent structures in spin-2 theory are separable. In spin-2-theory, the formerly irreducible metric field g of general relativity is split into two fields η and h; η is ontologically dependent on the material structures and h is an ontologically independent matter field.

Thus, general-relativistic spacetime (more precisely: the spacetime metric g) can be derived from matter structures. The dynamical spin-2 view provides an interpretation of general relativity that supports a form of spacetime non-fundamentalism. This in itself is an important finding since it refutes the worry that spacetime non-fundamentalism is incompatible with general relativity. However, the non-fundamentality of general-relativistic spacetime does not follow automatically, i.e., not directly from the physical theory and its formalism. The question is which interpretation is most convincing. While the non-fundamentalist interpretation of the dynamical spin-2 view of general relativity is a very good candidate for the best interpretation, I concede that the situation is not clear-cut yet. This is mainly because the formally symmetric dependence relation between g, η, and h also allows for a spacetime fundamentalist reading that, by standard evaluation criteria, such as explanatory strength and parsimony, does not score so poorly to disqualify unambiguously. After discussing further objections, I sum up that the standard fundamentalist interpretation and the non-fundamentalist dynamical spin-2 interpretation of general relativity are approximately on a par. Thus, there is a tie between spacetime fundamentalism and spacetime non-fundamentalism.

Only in the third step is spacetime fundamentalism rejected as untenable, and that within the overall context of physics – in particular taking into account arguments against semi-classical theories of gravity and proposals for a theory of quantum gravity. In Chapter 5 ("Spacetime in Quantum Gravity"), I therefore turn to the research programme of quantum gravity. To emphasise the general relevance of the research programme, I lay out in detail why physicists are searching for a theory of quantum gravity. I defend the standard arguments against semi-classical theories against the prevailing criticism in the philosophical literature. I then give an overview of some of the major candidate theories and examine their implications for the spacetime debate. Since the proposed theories differ widely, it cannot be said that theories of quantum gravity *per se* are opposed to any form of spacetime fundamentalism – contrary to what is usually claimed. Instead of ignoring "inconvenient" theories, as is often done in the literature, I propose, for the sake of robust metaphysical inferences, to look more closely at the very theory that seems particularly unfavourable to spacetime non-fundamentalism: quantum spin-2 theory.

I first present the foundations of quantum spin-2 theory, including its relation to classical spin-2 theory and general relativity. On this basis, I then propose, amongst other things, a new understanding of the equivalence principle. With respect to the fundamentality of spacetime, I argue that it is quantum spin-2 theory that breaks the tie between spacetime fundamentalism and spacetime non-fundamentalism: a dynamical interpretation of quantum spin-2 theory reveals spacetime (more precisely, the spacetime metric) as a non-fundamental structure. In short, this is because general relativity can be reduced to quantum spin-2 theory.

Now, to ensure the robustness of our metaphysical judgments (e.g., on how spacetime exists), it is advisable to take into account all theories which are compatible with the available empirical data as well as all tenable interpretations of these theories. Accordingly, the mere existence of the geometrical interpretations of classical general relativity and classical spin-2 theory, which are indeed compatible with the available empirical data, still interferes with concluding that metrical aspects of spacetime are non-fundamental. Moreover, the question arises whether future theories might rehabilitate spacetime fundamentalism. Therefore, in Chapter 6 ("Unification"), I conclude by proposing a solution to this problem via the unificatory practise of physics. In short, the idea is to constrain the ontological conclusions in the philosophy of spacetime by continuity conditions to other parts of physics, in particular elementary particle physics.

It is such conditions, I argue, that provide criteria for determining which of the various conflicting ontological commitments from the philosophy of spacetime are ultimately preferable: namely, those that also respect and are compatible with central insights from other areas of physics. Specifically, spacetime fundamentalism may be a tenable position when considering only a particular class of spacetime theories. It becomes untenable, though, if one includes, for example, the established insights of quantum physics.

For such a "unifying" argument to be convincing, the physical practise of unification should not itself be based on metatheoretical or metaphysical presuppositions. I argue that this is the case. The unificatory practise of physics does not depend on assumptions external to physics. Rather, unification is the internally obtained result of good scientific practise – virtually its by-product.

Finally, I propose concrete continuity conditions for the case study of spacetime. I emphasise that all theories compatible with the available empirical data have interpretations according to which metrical aspects of spacetime are non-fundamental, while only some theories can be interpreted as implying that metrical aspects of spacetime are fundamental. This suggests the inductive metaphysical inference that metrical aspects of spacetime are indeed non-fundamental.

2 Fundamentality

As this work is about the question of whether spacetime is fundamental according to relativity theory and theories of quantum gravity, I am mainly concerned with studying these theories and interpreting what they tell us about spacetime. The metaphysical conclusions that I draw are to be understood as the *results* of these concrete investigations of physics. However, it is important to clarify beforehand what it means to say that spacetime is or is not "fundamental". To do this rigorously requires an initial dive into metaphysics. This not only serves to illustrate, from the start, how my project relates to previous debates on spacetime, but additionally indicates connections to general issues in metaphysics as well.

I shall therefore begin by explicating "fundamentality" in general metaphysical terms in Section 2.1. I then specify the discussion to the case of spacetime in Section 2.2 by briefly reviewing the traditional debate on the ontological status of spacetime, traditionally referred to as the substantivalism–relationalism debate. In addition, connections are drawn to more recent issues regarding the frequently invoked "emergence of spacetime". As a side note, I point out that spatiotemporality should be considered to come in degrees.

2.1 What Is Fundamentality?

In contemporary metaphysics, fundamentality and related concepts are intensively debated. The notion is also frequently used in physics and the philosophy of physics. Often, however, it is not sufficiently clear what it is supposed to mean. Physicists and philosophers of physics usually do not pay much attention to how key metaphysical notions like "fundamentality" are to be understood precisely. The metaphysical debate on fundamentality is usually not registered in physics or the philosophy of physics.

Physicists typically employ the notion without further analysis of its content, i.e., as a primitive concept that is then applied to entities, equations, and theories alike. For instance, the fundamentals are what features in the fundamental Lagrangian (Weinberg & Witten, 1980, p. 59). Or,

DOI: 10.4324/9781003404149-2

the Schrödinger equation counts as fundamental. Or, theories of quantum gravity are viewed as fundamental (Kiefer, 2007). A weak criterion for classifying an equation as fundamental is that it is central to some theory. A stronger criterion additionally claims that there is a sense in which the theory is also fundamental. A theory is often viewed as fundamental because it provides the most general, conclusive, and accurate description of its subject matter. Hence, fundamentality is often linked to unification (e.g., Wald, 1984, p. 379; Weinberg, 1993).

If the notion is further analysed, it is typically understood in terms of (a broad notion of) composition. For example, particle physics seeks to find the "fundamental constituents of the universe" (Thomson, 2013, p. 1), implying that the fundamentals are not themselves constituted. The fundamental entities are uncomposed, elementary, or "structureless" (Halzen & Martin, 1984, p. 4). A proton is composed of, amongst others, quarks, but quarks are uncomposed. So, protons are non-fundamental and quarks are fundamental (Weinberg, 1977, p. 174). The fundamental entities are those that do not have any microstructure – they are simple and "boring" (Weinberg, 1993, pp. 45–46). Alternatively, physicists sometimes seem to argue that the fundamental entities are those that feature at the highest energy scale (corresponding to the smallest spatiotemporal scale) – at least, this is how we might understand a general sentiment behind research programmes like quantum gravity (besides issues of generality and unification). This can also be related to the "cooling" history of the universe (Halzen & Martin, 1984, p. 352): the universe started out from a dense high-energy state that then expanded and cooled down. So, what is present at the highest energy scale is fundamental in the sense that it is "first".

Philosophers of physics often use fundamentality talk with respect to a specific theory or theoretical framework.[1] They tend to employ a primitivist notion of fundamentality that conveys a minimal meaning of the term informed by how physicists use it. Unlike notions like emergence, fundamentality hardly receives analytic attention by philosophers of physics. Seminal papers, like Huggett and Wüthrich (2013), do not define the notion. Similarly, Knox (2013) does not deem the fundamental in need of further explication, but the emergent: "For our purposes here, by 'emergent', I'll simply mean 'non-fundamental', and leave to one side questions about what extra content is required for a full definition of 'emergence'" (Knox, 2013, p. 346). Crowther (2016), on the other hand, is "not sure what we should take 'fundamental' to mean" and therefore prefers to avoid the term (Crowther, 2016, pp. 55–56). Instead, she distinguishes different "levels based on their relative energy scales, using the phrasing micro and macro" (Crowther, 2016, p. 56).

Occasionally, however, further specifications are given. Then, the fundamental entities are, for instance, those that (1) are defined in the (kinematically possible) models of a theory (Read, 2019), (2) feature in (almost) all models of a theory or (almost) all solutions of some relevant equation (Bartels & Wohlfarth, 2014),[2] or (3) are "self-standing" and "autonomous" (Brown & Pooley, 2006, p. 84), "ontologically distinct and primitive" (Read et al., 2018, p. 18), or, in fact, "ontologically independent" (Read et al., 2018, p. 20) – a standard explication in contemporary analytic metaphysics (see below).

So, what does "fundamentality" mean? Depending on the specific context, notions typically used interchangeably with "fundamental" are "basic", "primitive", "irreducible", "underlying", "rock-bottom", "substantival", "independent", "prior", "autonomous", "distinct", "self-standing", "nonderivative", "elementary", "uncomposed", "unbuilt", "ungrounded", "primordial", "essential", "indispensable", and so on. The overall idea is that questions of fundamentality are questions of ontology – notably, not in the sense of a Quinean "What exists?", but rather a (Neo-)Aristotelian "What grounds what?" (see Schaffer, 2009, p. 347). The question whether some entity is fundamental is not merely a question whether this entity is on the list of what there is, but about investigating its *ontological status with respect to other entities*. The fundamental entities are *ontologically privileged* with respect to all other entities. They are *ontologically prior* to them – in either of two related senses: their being *ontologically independent* of other entities or their being what other entities *ontologically depend on*.

2.1.1   *Two Tentative Explications*

Hence, following Bennett (2017), a tentative explication of what it means for an entity to be fundamental uses the notion of an "unexplained explainer",[3] the two halves of which reflect the first and the second account of fundamentality that I shall present in a moment: a fundamental entity is *not explained* by anything else – which refers to its ontological independence – and a fundamental entity is in the set of entities which *explains* everything else – which refers to its being what other entities ontologically depend on.

Accordingly, we can ask, for example, whether metrical aspects of spacetime are indeed unexplained, and whether metrical aspects of spacetime indeed do explanatory work themselves – for example, with respect to the symmetry properties of matter fields – as the standard view has it.

Another illustration that is regularly invoked in metaphysics when it comes to explicating what fundamentality is about employs the *"all God would need to create" metaphor*,[4] according to which

> the fundamental entities are all and only those entities which God needs to create in order to make the world how it is. So if God wants to create a world *w*, the fundamental entities will be the entities necessary

and sufficient for God to create in order for her creation to count as a creation of w. Likewise, if she changes her mind and decides to create w^* instead, she will alter her creation by changing what fundamental entities she creates; she need change her creation only in fundamentals in order to make it a creation of w^* rather than w.

(Barnes, 2012, p. 876)

Consider a world w with only one complex – i.e., composed, derivative, or in any sense *built*[5] – object that depends on certain primitive constituents (Barnes, 2012). The idea is that the creation of w is accomplished by creating the primitive constituents of the complex object and arranging them properly; God would not have to additionally create the complex object on top of its properly arranged constituents – it comes for free. In other words, by *explicitly* creating the fundamental entities and properties (including their relations), God would have *implicitly* created the non-fundamental entities and properties as well. Arranging the constituents properly, i.e., determining their relations (e.g., laws of nature plus boundary conditions[6]), is obviously crucial to actually obtain w.

Phrased in this metaphor, the very question of the work at hand is whether for creating our world (or worlds that are sufficiently similar to our world) God would need to create spacetime (or certain spatiotemporal aspects) explicitly as part of the list of fundamental entities and properties, or whether spacetime is already created implicitly, for example, by creating the fundamental matter fields and determining their dynamics.

2.1.2 Absolute and Relative Fundamentality

One can also distinguish fundamentality *simpliciter* from the concept of one entity or property being *more* fundamental than another entity or property – Bennett (2017) dubs this *absolute* and *relative*[7] fundamentality, respectively:

We also say that some phenomena are more fundamental than – or exactly as fundamental as – others. This is relative fundamentality talk, and it cannot be replaced by absolute fundamentality talk. One thing can be more fundamental than another even though the former is not absolutely fundamental. For example, carbon atoms are more fundamental than I am, and the property being a carbon atom is more fundamental than the property being a homo sapiens. Yet neither carbon atoms nor the property being a carbon atom is absolutely fundamental. They are not part of the rock-bottom story of the world; there is a further explanation of their nature; they are built.

(Bennett, 2017, p. 138)

For Bennett, the relation of relative fundamentality is supposed to apply not only to entities which are in a concrete dependence relation with each other, but to any two entities:

> One thing can be more fundamental than another despite the first's not in any way building the second, and the second's not in any way depending on the first. A hydrogen atom in Phoenix is more fundamental than a water molecule in Ithaca, even though those particular entities stand in no building or dependence relations at all.
>
> (Bennett, 2017, p. 138)

This is why, according to Bennett, relative fundamentality is constrained by, but not identical with, building or dependence relations and, hence, involves a rather complex machinery that I shall not discuss here.[8] Roughly speaking, she understands relative fundamentality as *ontological priority* (Bennett, 2017, p. 137). Since I will only be concerned with relative fundamentality claims with respect to entities that *do* stand in ontological dependence relations, I shall simply use what she rejects as too specific an account of relative fundamentality as such: I shall take the asymmetric ontological dependence relations as relations of ontological priority – and *vice versa*. If *A* depends on *B*, then *B* is prior to *A*. That which builds is prior to what it builds. In particular, that which is ontologically reduced is less fundamental than that which it is ontologically reduced to. After all, I am concerned with the question of whether an allegedly absolutely fundamental structure, namely some spatiotemporal structure, can be shown to *not* be absolutely fundamental by ontologically reducing it to something else, namely matter field dynamics. Moreover, when the non-fundamentality of spatiotemporal structure is established, I am finished; I do not further inquire whether the more fundamental structure is the *most* fundamental structure, i.e., absolutely fundamental. Note also that all entities, properties, and concepts that I am concerned with in this work are typically considered perfect candidates for absolutely fundamental entities, properties, and concepts. So, for the sake of simplicity, I might occasionally assume that a property of concern in this work, say, a dynamical symmetry property of matter fields, is absolutely fundamental if it is more fundamental than some other property, say, a property of the spacetime metric. This is not meant as pre-empting future developments of physics, of course. I treat any absolute fundamentality claim as revisable – as is standard practise in science-based metaphysics.

2.1.3   A Clarification

Let me clarify that there are also questions of fundamentality regarding theoretical notions, concepts, or principles. For example, we can ask whether the equivalence principle of general relativity is fundamental. One may want to understand this as a question about what is part of the Quinean ideology (rather than ontology) of a theory (e.g., Pooley, 2013, Section 6).[9] I shall simply assume that in this case the analysis of fundamentality as, say, (ontological) independence is appropriately generalised to include corresponding notions of independence (e.g., conceptual independence).

2.1.4 *Three Accounts of Fundamentality*

In the following I give a brief and non-comprehensive presentation of how the above translates to three standard proposals for explicating fundamentality in metaphysics.[10] According to the first account, the fundamental entities and properties are those which are *ontologically independent* – Bennett (2017) advocates this account. On the second account, the fundamental is that which is *in the minimal complete set of entities or properties that determine everything else*, which amounts to a straightforward interpretation of the "all God would need to create" metaphor – McKenzie (2018) defends this. Third, fundamentality is not analysed but *primitive* – Sider (2011) and Wilson (2014) endorse this view. Generally, these accounts can come apart, as I shall indicate in due course, but for my purposes, all accounts are tenable, or so I shall argue. So one purpose of laying out the different notions of fundamentality in this chapter is to make plausible that they are all compatible with what I adopt as a rather permissive notion of fundamentality. That being said, I prefer to understand fundamentality in terms of (ontological) independence. I do not engage in the debate about which of these accounts might be considered best in general, though.

2.1.4.1 *Fundamentality as Ontological Independence*

Bennett (2017) gives an account of fundamentality in terms of what she dubs "building". Roughly speaking, questions of building are

> questions of what 'gives rise to', 'makes up', or 'generates' another, or, to switch directions, about what some phenomenon is 'based in', 'constructed from', or 'built out of'.
>
> (Bennett, 2017, p. 2)

Naturally, two distinct explications of fundamentality are available: either the fundamental is *unbuilt*, or the fundamental is what *builds*. The first notion appeals to a "downward" directed sense of fundamentality, while the second appeals to an "upward" directed sense of fundamentality (Bennett, 2017, p. 111). Importantly, the building relations are *directed*, i.e., *asymmetric* (or, equivalently, irreflexive and antisymmetric). Accordingly, "the claim is that nothing builds itself, and no two things mutually build each other" (Bennett, 2017, p. 33).

Bennett reads the upward directed "builds" aspect as referring to the fact that the fundamentals are *complete*: the fundamental entities constitute the complete set of entities which build everything else – this is the second account of fundamentality (see below). The downward directed "is unbuilt" aspect of fundamentality refers to the fact that the fundamentals are *independent*. Accordingly, "*x* is independent if and only if *x* is not built by anything" (Bennett, 2017, p. 105). Bennett continues that the independence aspect arguably precedes the completeness aspect of the analysis (Bennett, 2017, pp. 122–123):

> The independents are in the unique complete set *because they are independent*. They are in the unique complete set because nothing else builds them, not because they do any building work of their own.
>
> (Bennett, 2017, p. 123)

Ultimately, this is why she proposes to understand fundamentality as independence. What is more, this establishes a sense in which the two accounts of fundamentality may be explanatorily related – which in turn explains why we may often switch between the notions methodologically. For example, we may be able to show that some entity is (e.g., for explanatory reasons) indispensably part of the complete set – and then argue that *this is why* we should consider the entity to be ontologically independent.

Since building is not a single relation, but a class of different building relations (Bennett, 2017, pp. 106–107) – like (mereological) composition, constitution, and grounding (Bennett, 2017, pp. 8–13) – there are different notions of ontological independence: an entity may be ontologically independent in the sense that it is uncomposed, not constituted, or ungrounded, for example.

Rather than relying on one of Bennett's building relations and the respective notion of ontological independence, however, I shall turn to (some form of) ontological reduction by noting that ontological independence arguably implies *ontological irreducibility*. So, per contraposition, if an entity is ontologically *reducible*, then it is ontologically *dependent*, i.e., *non*-fundamental. A similar idea is, critically, summarised by Fine (Bennett, 2017, pp. 134–135):

> It is natural to understand the concept of fundamental reality in terms of the … concept of one thing being less fundamental than, or reducible to, another – the fundamental being whatever does not reduce to anything else (but to which other things will reduce).
>
> (Fine, 2001, p. 25)

Note, however, that Fine's remark misrepresents the valid logical relations, when we assume Bennett's notion of fundamentality as ontological independence: *A*'s being "less fundamental", i.e., non-fundamental, i.e., ontologically dependent, does *not* imply that *A* is ontologically reducible. And hence, per contraposition, *A*'s being ontologically irreducible does *not* imply that *A* is ontologically independent, i.e., fundamental. In general, something can be irreducible, but still ontologically *dependent* in a substantial building-type sense. For example, mental properties may be irreducible, but still ontologically dependent on physical properties (and hence non-fundamental), as non-reductive physicalism asserts (e.g., Rolffs, 2023). While it may be rather safe to assume that the entities under consideration in this work do not pose such problems (e.g., a matter field's being irreducible *does* imply its fundamentality), it is important to note that this work only needs to assume that ontological reducibility entails ontological dependence, which is uncontroversial.

Connecting ontological dependence to ontological reduction fits best with my approach to the question whether spacetime is fundamental, which I understand as the question whether spacetime – or at least certain aspects of spacetime – are ontologically reducible to non-spatiotemporal entities or properties. In other words, I seek to establish that certain aspects of spacetime "ontologically depend on" non-spatiotemporal structure in the sense that these aspects of spacetime are "derivative on" non-spatiotemporal structure – contrary to the received view.

2.1.4.2   *Fundamentality as Completeness or Determination*

Starting from the "all God would have to create" metaphor, one may alternatively propose to read fundamentality as *determination* (McKenzie, 2018, p. 58) or *completeness* (Bennett, 2017, pp. 107–111) – or, indeed, *complete determination*[11]:

> for what it [the metaphor] connotes is that, by making the fundamental, everything else was settled, taken care of – or, in other words, *determined*.
>
> (McKenzie, 2018, p. 58)

Or, as Bennett puts it:

> The basic idea is that the fundamental entities are not those for which nothing else accounts, but rather those that do the accounting – they are the things that account for everything.
>
> (Bennett, 2017, p. 107)

It is important to note that it is about everything else being determined *completely* – which is reflected by Bennett's denotation.

Fundamentality understood as independence lacks this aspect of completeness, McKenzie (2018) continues to argue[12]: if there are dependent entities that are not completely determined by that on which they depend, fixing which entities are independent does not fix everything else. Incomplete reductions are examples of such cases. Thus, only fundamentality analysed as complete determination does fully account for the "all God would have to create" metaphor.

Here is another way to make plausible that analysing fundamentality in terms of determination or completeness is indeed distinct from analysing fundamentality as independence:

> suppose building could, *per impossible*, hold either reflexively or in a circle. A world in which there was nothing but self-built entities, or nothing but a building circle, would be a world in which there is a complete set, yet no independent entities at all. (I owe this point to Jessica Wilson).
>
> (Bennett, 2017, p. 111)

For my purposes the two notions do not come apart in a relevant sense, because I am mostly concerned with cases of complete reducibility. Hence, I may occasionally switch from independence to determination talk for clarification but will typically use the independence notion. For establishing that metrical aspects of spacetime are non-fundamental, I dispute the ontological independence of metrical aspects of spacetime, i.e., I argue that these aspects are ontologically reducible, viz., ontologically depend on something else (alternatively: are determined by something else). I do not argue that the metric is non-fundamental because it does not determine something else. Even less do I argue that the metric is non-fundamental because it is not part of the minimal complete set that determines everything else. As a result, fundamentality as ontological independence better meets my purposes and better reflects my methodology.

With respect to the debate on the dynamical approach to general relativity in Section 3.5, however, both notions of fundamentality arguably do come apart in the following sense: according to the standard dynamical approach by Brown (2005), the metric field g is indeed dependent, but not fully determined by other posits. In Section 3.6, on the other hand, we encounter arguments on whether topological aspects of spacetime are *explanatorily indispensable*. Here, the fundamentality of topological properties is arguably debated in terms of completeness or determination.

2.1.4.3   *Fundamentality as a Primitive Notion*

Alternatively, some argue that fundamentality is a primitive notion that cannot or should not be analysed in terms of other concepts (see also Mc-Kenzie, 2022, p. 3):

> the fundamental should not be metaphysically characterized in nega-tive terms [like "ungrounded"; my remark] – or indeed, in any other terms. The fundamental is, well, *fundamental*: entities in a funda-mental base play a role analogous to axioms in a theory – they are basic, they are 'all God had to do, or create'. As such – again, like axioms in a theory – the fundamental should not be metaphysically defined in *any* other terms, whether these be positive or negative.
>
> (Wilson, 2014, p. 560)

In principle, I feel no need to exclude this way of thinking about funda-mentality – for my purposes, the same entities will count as fundamental on both the primitivist and the independence notion of fundamentality (the independent entities are precisely those entities that are in the minimal set of all that God would need to create). In fact, some may argue that this is the best notion for my purpose, as it avoids the troubles of defining in abstract and potentially obscure metaphysical terms what is a common and usually well-understood notion: being a fundamental physical entity or property. In my often bluntly speaking of "fundamentality" one may sense an acknowledgement of this. However, as argued above, there are concrete and meaningful explications at our disposal that reflect both the specific philosophical context of this work and its methodology. Hence, there is no good reason for preferring to posit fundamentality as an unex-plained primitive notion – especially, since also primitivists do inevitably give at least some tentative explication to convey the meaning of the no-tion in the first place.

2.1.5   *A Remark on Ontological Reduction*

Let me also briefly (and arguably crudely) comment on the notion of onto-logical reducibility and its relation to ontological dependence. Ontological reducibility is "typically associated with the idea that entities, proper-ties … at one level are 'nothing more than' a manifestation of entities, properties … at a lower level" (Morrison, 2006, p. 877). In other words, the lower-level properties completely determine the higher-level proper-ties. As I have pointed out above, fundamentality implies ontological ir-reducibility (but not *vice versa*). Hence, demonstrating that an entity or property is ontologically reducible is sufficient for demonstrating that it is non-fundamental (ignoring the problem of fixing the direction of reduction relations for now; see below). As the presentation above indicates, ontological dependence is a broader and generally weaker notion than ontological reducibility. For example, mental properties may ontologically depend on physical properties without being reducible to them, as non-reductive physicalists hold. In turn, if something is ontologically irreducible, it need not be ontologically independent, i.e., fundamental. So, fundamentality, i.e., ontological independence, is a stronger notion than ontological irreducibility. In this sense, my focusing on ontological reducibility *specifies* the dependence relation that I am concerned with.

A standard way to spell out ontological dependence so that it connects to ontological reducibility is in terms of *supervenience*. In short, "a set of properties A supervenes upon another set B just in case no two things can differ with respect to A-properties without also differing with respect to their B-properties. In slogan form, 'there cannot be an A-difference without a B-difference'" (McLaughlin & Bennett, 2018). In the spacetime context, for example, Norton (2008) uses the supervenience relation to characterise how the dynamical approach to special relativity conceives of the Minkowski metric, namely as supervening on matter field properties (see Section 3.5).

However, the supervenience relation alone is not sufficient for making the ontological dependence relation precise: "A supervenes on B" merely expresses "that certain patterns of property ... variation hold" (McLaughlin & Bennett, 2018, Section 3.7). But we would like to say, for example, that the Minkowski metric has its properties *in virtue of* the matter field properties; supervenience is not able to provide such explanations. It is here where standard notions of ontological reduction in terms of *derivation* or *functional realisation* are helpful, the latter of which has recently received some attention in the philosophy of spacetime literature.

The general concept of reduction in terms of derivation was famously proposed by Nagel (1961).[13] It roughly states that terms and relations in a target theory can be related to terms and relations in a base theory via bridge laws. With regard to at least some of the cases discussed here, this potentially complex scheme simplifies significantly, since the target theory is indeed shown to be deducible from the base theory. Reduction as derivation is applicable, for example, with respect to the reduction of general relativity to a classical spin-2 theory, or so I shall argue, with a bridge law that identifies the key terms (the reduction is secured by the identity $g = \eta + h$). I shall take it that this notion of reducibility obtains ontological import by noticing that the terms of the base theory explain the terms of the higher-level theory and reveal them as derivative. So, for example, η and h, which represent certain fundamental entities and properties, explain g, which then represents a derivative entity with derivative properties. It is important to note, though, that such arguments are less straightforward than just presented, since said identities are symmetric. Establishing a fixed direction for such relations requires further argumentation and may involve a non-literal reading of the identities (see Sections 4.2 and 4.3).

Functional realisation is another standard concept that can help to substantiate supervenience relations.[14] It is standardly used in the philosophy of mind, where it is usually assumed that we cannot deduce the higher-level mental concepts (e.g., pain) from the lower-level physical concepts (e.g., a certain neural activity), and where issues of multiple realisability arise. In broad strokes, the idea of functional realisation is that an entity or property A is functionally reducible to a non-empty set of distinct entities or properties B, if it is demonstrated that and how B takes over, i.e., "realises", the functional roles of A (see Kim, 1998). Here, a functional role is, roughly speaking, a relational redefinition of the entity or property in question. So for functionally reducing an entity or property, one first needs to define the entity or property that is to be reduced in terms of its functional role(s) – "in other words, a functionalist must establish (or, at least, assert) that there is nothing more to the concept in question than the particular role that the functionalist specifies" (Crowther et al., 2021, p. S222). Second, one needs to argue that the functional roles of the entity or property that is to be reduced are realised by some entity or property in the reducing theory and assert that, to paraphrase Crowther et al. (2021, p. S222), this entity which realises the functional role in that theory *just is* the property or entity that is to be reduced. In this way, functional realisation explains why a certain supervenience relation and, potentially, certain identities hold. Thereby functional reductions can complement other concepts, for example, reduction as derivation (Ney, 2008, Section 2a). With regard to the spacetime context, the concept of functional reduction can help to understand, for example, the status of chronogeometricity in Brown and Pooley's dynamical approach to general relativity and the sense in which some theories of quantum gravity like causal set theory may be able to recover classical continuous spacetime (Lam & Wüthrich, 2018).

The concept of functional reduction makes clear that ontological reductions can be incomplete. For example, metrical properties of spacetime may have functional roles that are realised by something else, but topological properties of spacetime may not. Then metrical properties can be considered completely ontologically reducible, but topological properties are irreducible. Spacetime is only ontologically reduced in part; in other words, the metric is non-fundamental, whereas the manifold may still be fundamental. Similarly, if only some of the properties of the metric field g are completely ontologically reducible, then the metric field is only ontologically reduced in part; in other words, the reducible property is non-fundamental, whereas the metric field g *without this property* may still be fundamental. We encounter such a case when discussing the standard dynamical approach to general relativity. As mentioned above, the two standard concepts of fundamentality, independence and determination, can come apart for incomplete reductions.

For emphasis, notice that I do *not* argue that only fundamental objects exist, nor that there is only one layer of reality. So even if *A* is completely ontologically reducible to *B*, this does *not* mean that *A* does not exist, i.e., that *A* is to be removed from the overall ontology. It only means that *A* is not part of the *fundamental* ontology; *A* can still exist as a derivative object. A table, for example, is ontologically reducible to certain fundamental quantum particles and their interactions. Thus, the table is non-fundamental. However, the table does still exist, namely as a derivative object. The table is not part of the fundamental ontology, but still part of the overall ontology.

2.2   Fundamentality of Spacetime

Let me move on to discussing the more specific issue of what "fundamentality" means in the philosophy of spacetime context.

To ask whether spacetime is fundamental is to ask about the ontological status of spacetime. According to Pooley, it means asking whether spacetime is "an entity in its own right" (Pooley, 2013, p. 522). Maudlin uses the same expression and also speaks of "independent existence" (Maudlin, 1993, p. 184). Both characterisations fit well with my preferred understanding of fundamentality as ontological independence. So, the question is whether spacetime is an ontologically independent entity. The traditional way of addressing this question – which originally was discussed for (Newtonian) *space*,[15] of course – is by asking whether spacetime is a *substantival* "container", or whether spacetime is "nothing but" the spatiotemporal *relations* between material objects.[16] This is the *substantivalism–relationalism debate*.[17] Although the debate is far from being settled and went back and forth in the past, substantivalism is arguably the mainstream position, in particular with respect to the three most important spacetime theories, namely Newtonian mechanics, special relativity and general relativity. In order to be able to present this immensely complex debate in a nutshell, I shall make some (further) simplifications in the following, not all of which – my apologies to the reader – are explicitly flagged. For instance, I do not attempt to give a historically accurate presentation when occasionally referring to the main historical opponents, Gottfried Wilhelm Leibniz and Isaac Newton.

2.2.1   *The Traditional Substantivalism–Relationalism Debate*

Roughly speaking, substantivalists hold that spacetime is on the list of the fundamental entities that "God would need to create",[18] whereas relationalists hold that it is not, but is derivative on material entities and their properties and relations. According to substantivalists, spacetime is ontologically independent. According to relationalists, spacetime ontologically depends on material entities and their properties and relations. Newton is standardly understood to have provided a defence of substantivalism (Pooley, 2013, p. 523) against which Leibniz argued in the *Leibniz–Clarke debate* – himself advocating relationalism.

2.2.1.1 *Newton's Bucket-Type Arguments*

Both camps have put forward various arguments for their position. Most importantly, Newton can be viewed to have argued for the physical – and hence ontological – significance of absolute space by what is now dubbed the *Newton's bucket argument* which has an actual experimental set-up; a related thought experiment depicted in Figure 2.1 is known as *Newton's rotating globes*.[19] In short, Newton argues that absolute space and time need to be assumed, because there are certain observable dynamical phenomena (inertial effects) for rotating systems which cannot be explained by the relative motion of the bodies involved; their relative motion is simply "not systematically correlated" (Maudlin, 1993, p. 184) to these phenomena. Essentially, this is an argument from explanatory indispensability that, in a methodological perspective, appeals to a notion of fundamentality as completeness: absolute space is fundamental, i.e., ontologically independent, because it is indispensable for explaining basic observations, it is indispensable from the minimal complete set that God would need to create. To counter this argument, one needs to show that absolute space is, in fact, dispensable, or that absolute space ontologically reduces to something else.

Newton's bucket-type arguments are usually understood to have some force against the relationalist, but they are not considered to banish the

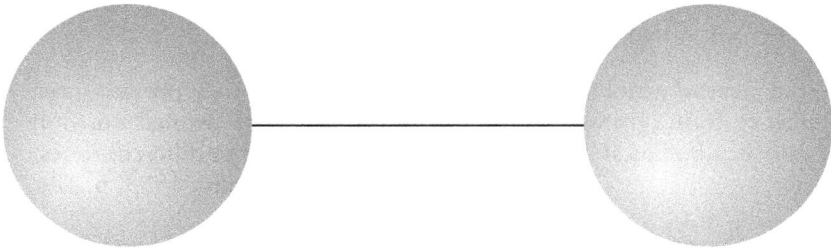

Figure 2.1 Newton's rotating globes thought experiment. The thought experiment considers an otherwise empty world with two globes that are connected by a rope.

position altogether. Here is why. Newton essentially argues that the relationalist does not have the explanatory resources to explain inertial effects. One obvious way to counter the argumentation is to show that she actually has: Mach (1919) famously points out that there are massive bodies available that can serve as reference for explaining Newton's bucket experiment in terms of relative motion (e.g., Earth or the fixed stars). However, a reply to Newton's globes thought experiment cannot proceed along those lines, since there simply is no additional body. As a result, the only option for the relationalist seems to lie in rejecting the thought experiment altogether – for example, on the basis of our intuitions' being unreliable in such highly hypothetical scenarios – and demand engagement with the *"actual facts"* (Mach, 1919, p. 229).[20] But what if the relationalist accepts the set-up of the argument?

First, here is the thought experiment in brief (following Maudlin, 1993): consider a world with only two globes connected by a stretched rope as depicted in Figure 2.1. The question is whether the following two states of the system are empirically distinct: the globes being at rest and the globes rotating with respect to their centre of mass. The crucial intuition that Newton's thought experiment appeals to is that the rotation results in a measurable tension of the rope. Accordingly, the two states are empirically distinct. The substantivalist has the explanatory resources to accommodate this: the two states are distinct because the two globes are either at rest or rotate with respect to absolute space. The relationalist, however, only has spatial distances between material bodies at her disposal, and thus seems unable to explain why the two states should count as distinct, and what should have caused the rope's tension. Thus, it seems that the thought experiment (if accepted) refutes relationalism.

But there is an important caveat to this. The notion of absolute three-dimensional space implicitly depends on its well-defined embedding in four-dimensional "space*time*" (not in the relativistic sense; see also endnote 15). As Maudlin spells out in detail, the decisive ingredient for being able to tell apart the two scenarios (the globes at rest, and the globes rotating) is that

> the parts of absolute space endure through time, they provide a reference frame by which distance relations can be defined not only between bodies at a time but also between bodies at different times.
>
> (Maudlin, 1993, p. 186)

It is not the posit of absolute space as an ontologically independent entity itself that brings about the required explanatory resources, but the fact that it provides "a means of extending the domain of the distance relation to include pairs of nonsimultaneous events" (Maudlin, 1993, p. 187).

This draws attention to the fact that Newton's bucket-type arguments – even when accepted as they are – do not tell against relationalism *per se*, but rather make explicit that the set of relations used must be sufficiently rich ontologically (Maudlin, 1993, p. 187). In particular, an apt set of relations needs to include distance relations between material events not only at the same time, but also at different times. This essentially defines a "'true' speed" (Huggett, 2006, p. 44) for the material body under consideration (in relation to its past position). The ontologically poor version of only having temporal relations plus spatial relations between simultaneous events is insufficient and indeed refuted, according to Maudlin.[21]

Before I turn to the two central arguments advanced by Leibniz against Newton's absolute space, notice that while this enriched version of relationalism might seem unnatural with respect to classical mechanics, such a version of relationalism is perfectly natural with respect to special and general relativity (Maudlin, 1993).

2.2.1.2 *Leibniz Shift Arguments*

Essentially, Leibniz's main arguments against Newton concern what Leibniz deems an unacceptable metaphysical stance[22]: having space and time as ontologically independent entities in addition to material bodies, the substantivalist has to embrace the metaphysical possibility of there being empirically indistinguishable worlds which nevertheless are ontologically distinct; in particular, "the whole material world could be displaced or set in motion in space without any consequent change in appearance" (Maudlin, 1993, p. 184). These make the two *Leibniz shift arguments*[23]: the *static* and the *kinematic shift argument*.

Here are the two arguments in more detail (following Maudlin, 1993; Pooley, 2013). Regarding the static shift argument, consider a Newtonian world w where material bodies are placed in a substantival "container" space. Leibniz then asks us to compare w to a world w' which is constructed from w by translating ('shifting') all material bodies, say, two meters to the east.[24] Leibniz argues that there is no *empirical* difference between w and w'. As a matter of fact, Newton's own theory only considers the bodies' relative distances, not their absolute positions; i.e., the theory treats all worlds that are related by a static Leibniz shift as identical – as does the relationalist. The substantivalist nevertheless maintains that w' is *ontologically* distinct from w, since the material bodies occupy different absolute positions. What is one and the same world for both the relationalist and, more importantly, Newtonian theory, for the substantivalist is infinitely many, ontologically distinct, but empirically indistinguishable worlds – arguably this does not conform to common metaphysical practise of not making ontological distinctions without good reason.

The kinematic shift argument is constructed analogously, but there are some notable conceptual differences which I comment on in a moment. The argument is based on the fact that Newtonian theory is invariant under Galilean transformations. Consider once again a Newtonian world w where material bodies are placed in a substantival "container" space. Compare w to a world w' which is constructed from w by adding a constant velocity to the absolute velocities of all material bodies at all times; say, five meters per second in westward direction. According to the substantivalist, w' is *ontologically* distinct from w, since the material bodies in w' collectively move differently relative to absolute space than the material bodies in w. But, again, there is no *empirical* difference between w and w' that underwrites this ontological distinction by the substantivalist: "by construction, the histories of relative distances between material objects in each world are exactly the same" (Pooley, 2013, p. 530) – which is all that matters according to the theory. Newtonian theory treats all worlds related by a kinematic Leibniz shift as identical; to paraphrase Maudlin (1993, p. 189): uniform absolute motion does not produce any observable dynamical consequences, only absolute acceleration does. The kinematic shift argument tells us that the substantivalist is committed to the material universe having an absolute velocity that is, in principle, empirically inaccessible. As Maudlin (1993, p. 192) and Pooley (2013, p. 530) stress, postulating empirically inaccessible physical facts may not be prohibited, but one should always prefer a theory that does without them.

Thus, both the static and the kinematic shift argument are said to demonstrate that the substantivalist accepts empirically indeterminable matters of fact as part of her ontology. Against this, Maudlin highlights the following subtlety. The two shift arguments, Maudlin argues, can be shown to operate by different rationales: the static shift argument presents a counterfactual situation and appeals to the principle of sufficient reason, whereas the kinematic shift argument is applicable to the actual universe and appeals to the principle of the identity of indiscernibles.[25] This casts doubt on the standard conclusion: it is only the kinematic shift argument that "gives rise to the existence of empirically undeterminable matters of fact" (Maudlin, 1993, p. 185).[26]

These subtle differences translate to distinct strategies with respect to the substantivalist's response to the shift arguments. On the one hand, regarding the static shift argument, the substantivalist essentially needs to do away with her haecceitistic approach to individuating space(time) points – indeed, merely maintaining that space(time) is fundamental in our *actual* world, substantivalism is not bound to space(time) points being also transworld-identical (Martens, 2019, p. 4). This can be achieved by adopting a version of *sophisticated substantivalism*,[27] which was originally proposed to counter attacks against substantivalism by the so-called *hole argument* – a static-shift-type argument against substantivalism in the context of general relativity[28]; the general idea is to individuate spacetime points by their metrical properties, i.e., to treat metric structure as essential to spacetime.

On the other hand, regarding the kinematic shift argument, the substantivalist needs to change the spacetime structure that she posits from a Newtonian to a Neo-Newtonian (or Galilean) spacetime structure to adopt to the dynamical structure of Newtonian theory.[29] Thus, conversely to the relationalist charged with bucket-type arguments (which revealed that she postulated an ontology too sparse to respect the dynamical symmetries), the substantivalist responds to the kinematic shift argument by committing to *less* spatiotemporal structure, since she started out by admitting too much of it.

So the general issue concerns a mismatch between the postulated spacetime structure and the theory's dynamical structure – i.e., between the *spacetime symmetry group*, which is a group of transformations "that *preserve[s] spatiotemporal structure* (as encoded in coordinate systems)" (Pooley, 2013, p. 528), and the *dynamical symmetry group*, which "*preserves the form of the equations* that express the dynamical laws" (Pooley, 2013, p. 529)[30]:

> the Galilean covariance of Newtonian mechanics tells against both substantival space and the most obvious relationalist alternative.
> (Pooley, 2013, p. 523)

This draws attention to the question whether the spatiotemporal and dynamical symmetries are to be considered as independent of each other, or whether there is any (e.g., explanatory) relation between them. Thus, it raises questions of fundamentality. Here this book enters the debate, as I shall develop further in Chapter 3.

To summarise, both substantivalists and relationalists can successfully resist the standard counter arguments by adjusting their ontological posits. In particular, the Newton's bucket-type arguments and the Leibniz shift arguments do not refute relationalism or substantivalism altogether, but constrain what the positions have to posit.

2.2.2 *Not an Issue of Realism*

While I have generally couched the debate as being about issues of fundamentality, some of the above formulations remind us that there is a sense in which the traditional debate as well as standard modern expositions are not originally about fundamentality claims, but about existence claims:

> The traditional ... debate is about whether space ... exists. The substantivalist says that it does. The relationalist says that it doesn't. According to the relationalist, all that exists, in the physical world,

are material bodies related to one another spatiotemporally; there is no further thing in which these bodies are located.

(North, 2018, p. 3)

Against this, I agree with North that the content of the debate is better formulated in terms of questions of fundamentality (which she understands in terms of some grounding-type notion):

Both views can countenance, or believe in the existence of, spatiotemporal structure. ... The views differ on what underlies this structure. Essentially, the substantivalist says that spatiotemporal structure is fundamental to the physical world, whereas the relationalist says that it arises from the relations between and properties of material bodies.

(North, 2018, p. 13)

This understanding of the substantivalism–relationalism debate is not the prevalent view in the philosophy of physics literature (e.g., Maudlin, 1993; Nerlich, 1994; Norton, 2008), as North acknowledges; the recent focus on issues of fundamentality in metaphysics, however, has arguably spawned a shift in philosophy of physics as well.[31] For instance, Martens argues that an existence claim about spacetime is already implied by a fundamentality claim about spacetime. So, regarding substantivalism, the existence claim is redundant. Furthermore, Martens continues, modern relationalism typically holds that spacetime is derivable from spatiotemporal properties of matter fields. But if the spatiotemporal properties are real, then spacetime is as well, "it is just not fundamental" (Martens, 2019, p. 11):

As an analogy, consider a chair from the perspective of a particle physicist. The particle physicist would not deny that the chair is real, but would deny that it is fundamental. The chair is made up of equally real, but more fundamental elementary particles (or fields, or strings, etc.).

(Martens, 2019, p. 11)

Phrasing the debate in terms of fundamentality is therefore preferable: it highlights most clearly where the two positions disagree. Thereby, this understanding of the debate is able to diminish – or even dispel – recurring worries about "whether there is any clear-cut dispute between the two sides" at all (North, 2018, p. 128).[32]

To summarise, the traditional debate on space(time) centres on the question of "What exists?", i.e., on "What is part of our ontology?" In contrast, the modern debate on spacetime centres on the question of "What is fundamental?", i.e., on "What is ontologically independent?"; granted spacetime is in the ontology, the question that remains to be answered is whether it is part of the *fundamental ontology*, i.e., in the subset of the ontologically independent entities.[33] Both camps agree that spacetime is in the ontology, but they disagree on its precise ontological status. They disagree whether spacetime is ontologically independent, i.e., fundamental. Consider again the analogy: the chair exists, i.e., is part of the ontology, but it is non-fundamental, i.e., it is not part of the fundamental ontology.

Note that I refrain from engaging in the debate on whether an entity's non-fundamentality implies its non-existence, as some metaphysicians argue (see McKenzie, 2018). Relatedly, I shall ignore that some metaphysicians argue that to be able to debate an entity's fundamentality, one must first establish whether it actually exists (see McKenzie, 2018). I shall simply assume that the material world does exist – as McKenzie puts it: "clearly *something* exists" (McKenzie, 2018, p. 61) – and that the entities posited by our best physical theories are good candidates for that "something" (because I do think that our best physical theories refer at least approximately). Even granted that there might be nothing rather than something, we can still point out relations of fundamentality with respect to our theoretical representation of the world, which is precisely what one investigates as a philosopher of physics anyway. One may take these remarks as an indication that I tend to think of fundamentality and existence claims as (largely) orthogonal issues. Conceptually, fundamentality claims always concern the *whole* structure of the world: the claim that some entity is independent is a claim about the dependence relations of that entity to *all* other entities. On the other hand, claims about the reality of some entity are strictly individual. Being real just means fulfilling a list of intrinsic criteria.

2.2.3 *Degrees of Spatiotemporality*

So, the question that remains to be answered is whether spacetime is fundamental. More precisely, the question is *which* spatiotemporal structure, if any, is fundamental. The substantivalism–relationalism debate is implicitly about what exactly spacetime is. According to the received view, it is the *pair* of manifold and metric structure. This already suggests that there are *different aspects* to spacetime or spatiotemporality (Jaksland & Salimkhani, 2023) – e.g., topological and metrical aspects. Similarly, recent functionalist approaches distinguish different functional roles of spacetime.[34] Hence, it might be the case that some aspects of spacetime are fundamental, while others are not.

Moreover, this draws attention to the fact that the verdict that spacetime is fundamental (or not) is ambiguous: "spacetime" can mean various things. One needs to specify which spatiotemporal aspects are subsumed by the notion.[35] In joint work, Rasmus Jaksland and I propose that there are the following five conceptions of spacetime (Jaksland & Salimkhani, 2023): (1) on a *thin* conception of spacetime, any spatiotemporal aspect is individually sufficient for spacetime; (2) a *thick* conception includes all spatiotemporal aspects as individually necessary; (3) on a *specified necessary* conception, specified spatiotemporal aspects are individually necessary for spacetime; (4) on a *specified sufficient* conception, specified spatiotemporal aspects are individually sufficient for spacetime; finally, (5) the *cluster* conception à la Baker (2021) views no spatiotemporal aspect as necessary, but certain clusters of them as sufficient for spacetime. In this work, I shall not adopt any general conception of spacetime, but simply specify the spatiotemporal aspects I consider when necessary: mostly metric structure, occasionally chronogeometricity, and topological structure in Section 3.6; but in much of the book, I'm afraid, I do not heed my own call for specificity and use loose spacetime talk.

The above also brings up the question of what should count as a spatiotemporal structure in the first place. Is every aspect of spacetime on its own to be considered a *spatiotemporal* structure? Is the manifold spatiotemporal, for example? Or is there some more substantial criterion to be fulfilled for something being classified as "spatiotemporal"?[36]

In particular, analysing spacetime in terms of spatiotemporal aspects suggests that *spatiotemporality comes in degrees* – a thought that can already be found in Huggett and Wüthrich (2013), Le Bihan and Linnemann (2019), and Linnemann (2021). Some spatiotemporal structures may be viewed as richer or *more spatiotemporal* than others: a spacetime that has both topological and metrical properties can be seen as more spatiotemporal than one that has only topological properties; the pair of manifold and metric is more spatiotemporal than the manifold alone. Following this line of thought, one may attempt to order the different degrees of spatiotemporality to obtain something like a *hierarchy of spatiotemporal structure*. Note, however, that such a project is non-trivial due to various interdependencies between the different mathematical structures representing the different spatiotemporal aspects, including manifold and metric.[37] Moreover, it is arguably unclear how the different aspects of spatiotemporality, say, dimensionality, continuity, and spacetime split, compare with respect to the degree of spatiotemporality they imply. What is more spatiotemporal, a discrete structure of a specific (finite) dimension or a continuous structure with infinite dimensions?

Accordingly, having different aspects of spacetime does not imply having a well-ordered hierarchy. Nevertheless, one can point to ways in which some posited spatiotemporal structure is more or less general. For example, take what we standardly consider to represent spacetime in general relativity: the pair $\langle M, g \rangle$ of manifold M and metric g. This defines a *metric structure* which gives the spatiotemporal distance between any two events. An arguably less structured (or more generic) pair $\langle M, g_i \rangle$ replaces the metric g by a *conformal metric* g_i, i.e., by an equivalence class of metrics g_i, with $g_i = \xi^2 g_j$ $(i \neq j)$, where ξ is some real-valued smooth function on M. Conformal structure is less structured – and hence less spatiotemporal – because it only defines a scale-invariant *causal structure*. The metrics g_i are equivalent up to a scale factor. Accordingly, conformal structure only fixes which events are causally connected, but cannot provide further information about distances; for the class of Lorentzian metrics, this essentially amounts to introducing a light cone structure.

Descending further, we may even dispense with causal structure and only retain a *space–time split* by using the manifold M alongside some *vector field* V^μ, for example; the vector field introduces an orientation and thereby gives rise to a space–time split. Le Bihan and Linnemann (2019) and Linnemann (2021) argue that such a space–time split is essential for something being classified as "spatiotemporal". Hence, it would mark the lowest degree of spatiotemporality. Consequently, *topological structure*, i.e., the manifold M *without* additional posits, is *not* considered spatiotemporal. It does at least seem imprecise to label a structure "spatiotemporal" that cannot even express the fact that spatiotemporality is literally about both spatial *and* temporal aspects.

Especially in light of the dynamical approach and spacetime theories with two metric fields (so-called bimetric theories), another aspect is put in the spotlight: a metric structure *can be surveyed or not surveyed* by material rods and clocks. If it is surveyed, the metric is said to have an additional property: *chronogeometricity*. Having chronogeometricity is what makes *a* metric, *the* metric (which is physically significant and therefore operationally accessible). Chronogeometricity can be viewed to add another degree of spatiotemporality.

Further degrees of spatiotemporality are discussed in the literature. For example, Huggett (2017) studies the property of having a definite radius, and Martens (2019) takes "spacetime" and "spatiotemporal" to refer only to that which has *all* the properties we typically attribute to phenomenal spacetime (including, for example, its four-dimensionality).

2.2.4   *Spacetime Emergentism*

Martens' narrow notion of spacetime (and matter alike) has further consequences: it suggests promoting the traditionally dichotomous debate – where spacetime is either fundamental, or derivative on *matter* specifically – to a trilemma. If spacetime is only that which is "fully spatiotemporal" (in a sense yet to be specified) and matter is only that which is "fully material" (in a sense yet to be specified), then the fully spatiotemporal may either be fundamental, or derivative on the fully material, *or arise from other fundamental degrees of freedom that are neither fully material nor fully spatiotemporal* (Martens, 2019); "[t]he point is that the other degrees of freedom do not have the complete set of properties of either spacetime or matter, and hence belong to a third category" (Martens, 2019, p. 11). Relatedly, Martens and Lehmkuhl (2020a, 2020b) investigate in detail the tenability of a strict conceptual distinction between matter and spacetime on the basis of dark matter theories. In fact, another interesting case for studying the matter–spacetime distinction is presented in Chapter 4: the spin-2 field.

Martens dubs the third option *spacetime emergentism*.[38] Especially with respect to theories of quantum gravity, spacetime emergentism has become the focus of attention: most theories of quantum gravity seem to pose a problem for spacetime substantivalism, as they, in one way or another, do away with spacetime, or aspects thereof, at the fundamental level, but do not simply render spacetime as derivative on what we typically call matter. Martens refers to causal sets (in causal set theory) and spin networks (in loop quantum gravity) as examples for candidate underlying structures that are neither material nor spatiotemporal in the full sense. Conceptually speaking, this also amounts to a shift from subtle philosophical arguments to speculative arguments from physics (Martens, 2019).

I agree with Martens regarding his general analysis – in particular with respect to the importance of reconsidering the substantivalism–relationalism debate in the light of quantum gravity research. Still, I shall not adopt his threefold distinction because I do not share his narrow notions of spacetime and matter, which provide the logical space for a third concept. Stripping away some aspects of Martens' conception of spacetime certainly results in less spatiotemporality, as I have argued myself, but this does not imply a loss of spatiotemporality *per se*. The same applies to matter. According to a common narrow conception of matter, matter is massive and fermionic; for example, electrons and quarks count as matter, but photons do not. In contrast, I shall also regard the photon and other bosons as matter. This is licenced by physics. Relativity theory teaches us that mass is just "a manifestation of energy and momentum" (Carroll, 2004, pp. 49–50). In addition, consider the fact that the mass of classical macroscopic matter is *mostly* due to the energy deposited in bosonic gluons in protons and neutrons, and not due to fermionic quarks and electrons. Since this work is concerned with underlying structure that is sufficiently alike matter, it seems even less urgent to posit a third category. Accordingly, I shall stick to the common distinction of there being two categories of candidate fundamental structure, spatiotemporal structure and material structure, which both come in degrees. Thus, on my account, the spin-2 field in Chapters 4 and 5 is matter. This is supported by the fact that it carries energy, self-interacts, can be zero, and does not have Lorentzian signature, for example.

Still, in order to make explicit the remaining discontinuities of my project with the traditional substantivalism–relationalism debate, I, too, shall adjust my terminology. In light of this chapter, I generally refer to substantivalism as *spacetime fundamentalism*, and to the opposing position as *spacetime constructivism*. The latter denotes all those positions that do not accept spacetime as fundamental, i.e., relationalist and emergentist proposals alike. Thus, the substantivalism–relationalism debate becomes the fundamentalism–constructivism debate that this work centres on.

One last remark: in the philosophy of spacetime, investigations into the fundamentality of spatiotemporal aspects are often accompanied by or couched in terms of "emergence". Emergence talk prevails especially in the philosophy of quantum gravity, but is also viewed to receive support from certain interpretations of non-relativistic quantum mechanics, namely wave function realism (Wüthrich, 2018, p. 316). It is important to note that these debates do not refer to traditional philosophical accounts of emergence. In brief, accounts of "emergent spacetime" are generally *reductive* accounts of spacetime, where "emergent" is simply supposed to stress a certain autonomy and novelty of that which is reduced (Crowther, 2016).[39] I shall therefore take it that emergence talk merely highlights the occurrence of some reducible, but robust, largely autonomous, and, in a sense, novel behaviour or new property – very much like certain macroscopic properties of a table (e.g., its hardness) may be said to "emerge" from underlying microstructure. Obviously, the sole fact that a new property appears does not make the case against reductivism, but is only why we ask for an explanation of that property in the first place.[40]

Notes

1 Arguably, this is in continuation of the widespread Quinean approach to ontology. However, given that in metaphysics the fundamentality of some entity is standardly conceptualised as involving relations to all other entities of a world, this approach might be inadequate and an impediment to debates on fundamentality (as long as the physical world is not described by a single theory).

2 Such modal notions of fundamentality that refer to (some type of) generality, i.e., the presence of an entity in (almost) all models of a theory, might deprive us from being able to provide a satisfactory analysis of fundamentality with respect to a *particular* world alone, like our actual world; there is a sense in which each solution or model of a theory corresponds to a possible world – note, however, that one may argue that various solutions are needed to describe (different parts of) a particular world. The idea of such a modal conception of fundamentality is that an entity's presence in all models indicates that we cannot do without it – hence, the entity is fundamental: it is indispensable. However,

an entity might simply be fundamental in this very sense (i.e., indispensability) in specific worlds (due to the internal structure of such worlds) – *without* being indispensable in all possible worlds described by the theory. What is fundamental (in a particular world) and what is so necessarily (in all worlds) are distinct questions. For example, a fundamental entity of our world, which is described by some model of a theory, might not be instantiated in other worlds (i.e., other models of that theory).

This is to be distinguished from the fact that fundamentality is nevertheless always about necessitation, i.e., other possible worlds, in the following sense: Asserting that A is non-fundamental and grounded in α means to assert a necessitation relation between α and A: $\square(\alpha \to A)$, i.e., all worlds with α are also worlds with A. But, of course, there might still be worlds without α, but with A. In some worlds, A might be grounded differently, say, in a. For example, a might ground α. In such worlds, α is non-fundamental (i.e., fundamentality of α cannot be analysed simply in terms of α being present in all worlds). Notably, modal necessitation is typically viewed as not sufficient for grounding relations, which many take as hyperintensional (McKenzie, 2018). I thank Matthias Rolffs for helpful comments.

3 See Bennett (2017, p. 111). Instead of explanation talk, Barnes (2012) uses truth maker talk. Moreover, Bennett (2017) may arguably phrase this in terms of an "unbuilt builder".

4 See, for example, Schaffer (2009), Barnes (2012), and McKenzie (2018).

5 See Bennett (2017). The reader may understand this notion in the broad sense of "that which builds, generates or brings about something else". In particular, one is not restricted to mereological composition. Bennett even takes causal relations as building relations.

6 This does not mean that laws of nature are fundamental or primitive à la Maudlin (2007). One may still defend a Humean regularity account, for example, according to which laws of nature merely express certain regularities in the arrangement of the constituents.

7 As a side note, taking the concept of relative fundamentality seriously does not automatically involve a *hierarchical* view of fundamentality with several levels, as opposed to a *fundamentalist* view of fundamentality with only the two levels of the fundamental and the derivative. Brief expositions of the hierarchical and fundamentalist view can be found in Barnes (2012) and McKenzie (2018), for example. Moreover, talk about absolute fundamentality does not establish, of course, that there actually is an absolutely fundamental basis. Cartwright (1999) rejects fundamentalism on general methodological grounds, McKenzie (2011; 2017; 2022) argues that it is at least conceivable that good candidates of "fundamental theories" actually have *anti-fundamentalist* consequences. Note, however, that certain no-go theorems, which are derivable in these very frameworks, might push against this. For example, Weinberg and Witten's (1980) results put severe constraints on the non-fundamentality of the graviton.

8 The reader is referred to Bennett (2017), Chapter 6.

9 Quine (1951) distinguishes the ontological commitments of a theory from its ideological commitments (Bricker, 2016): "Given a theory, one philosophically interesting aspect of it into which we can inquire is its ontology: what entities are the variables of quantification to range over if the theory is to hold true? Another no less important aspect into which we can inquire is its *ideology* (this seems the inevitable word, despite unwanted connotations): what ideas can be expressed in it?" (Quine, 1951, p. 14). For example, general relativity

is committed to the existence of material rods and clocks (see Brown, 2005; Read, 2020), although the respective concepts are not expressible in the theory itself. Accordingly, ontological commitments and ideological commitments "are largely independent of one another ... In the other direction, a theory may be ideologically committed to relational concepts that add structure to the entities posited by the theory without adding new kinds of entity to the ontology" (Bricker, 2016, Section 1.1). Bricker points out that some consider the ideological commitments of a theory as crucial with respect to the question of theory choice (see Sider, 2011, pp. 12–15) who also stresses that "[a] theory's ideology is as much a part of its worldly content as its ontology" (Sider, 2011, p. 13). I shall not make explicit use of the ontology–ideology distinction. If a theory adopts, say, Minkowski spacetime, I shall simply view Minkowski spacetime as part of the theory's *ontology* rather than ideology – *pace* Sider.

10 In addition, there are accounts that understand fundamentality in terms of grounding (e.g., North, 2018), or according to Lewis's notion of naturalness (see Bennett, 2017, Section 5.7).

11 Besides McKenzie (2018), for example Paul (2012) and Barnes (2012) can be read to adopt this notion of fundamentality (Bennett, 2017). Note, however, that Barnes explicitly writes that she takes fundamentality as "a metaphysical primitive, and as such I will not attempt to give a definition of it" (Barnes, 2012, p. 876).

12 I dispense with her knowledge talk here.

13 See Ney (2008), van Riel and Van Gulick (2019).

14 The following is again solely meant to indicate how one might conceive of ontological reduction. Notably, one can take functional realisation to motivate what has recently been dubbed *spacetime functionalism*. Functionalist approaches to spacetime theories are investigated by a number of philosophers (see endnote 34), but I do not pursue or endorse such a view.

15 Actually, there is a sense in which Newton had in mind a space *and time* structure as well (albeit non-unified). In his *Scholium*, Newton defines absolute space, and with it absolute position, as that which is "immovable" (Newton, 1846, p. 77), i.e., that which does not change its position *in time* (see also Stein, 1967): "Newton's theory of absolute space concerns the structure, not of space only, but of space-time. It implies not that 'space is absolute' – whatever that might mean in Leibnizian or any other metaphysical terminology – but that space is connected with time in such a way that states of motion are well-defined" (DiSalle, 2006, p. 26). In fact, this aspect of four-dimensionality is crucial for properly understanding the force of Newton's bucket-type arguments against the relationalist (see Maudlin, 1993).

16 Arguably, this is already a simplification. The "container" view of spacetime – where the material objects are situated in the substantival spacetime "container" – is one of at least two general ways in which substantivalism may be understood (see Maudlin, 1993). The other, more radical position, is to conceive of spacetime as substantival *and* as the bearer of all properties including those predicated to material objects – a position known as super-substantivalism (e.g., Lehmkuhl, 2018). By contrast, substantivalism is compatible with matter also being a substance – or, more accurately, with certain matter fields also being substances. As this work is only concerned with the question whether spacetime is fundamental, I can safely ignore the more radical variant of substantivalism that additionally claims that matter fields are non-fundamental. However, I shall come back to the underlying issue in Section 2.2.4.

17 See Pooley (2013) for a modern introduction.

18 In Leibniz's words, "supposing space to be something in itself, besides the order of bodies among themselves" (paragraph 5 in Leibniz's third letter to Clarke; Leibniz & Clarke, 2000, p. 15).

19 It is important to note that the precise content and purpose of the Newton's bucket argument has been debated extensively. According to the traditional view, Newton advanced the argument in his *Scholium* of the *Principia* in support of substantivalism and absolute space. This view has been challenged and rejected, however, by a number of scholars – for example, Laymon (1978), Rynasiewicz (1995a; 1995b; 2014), DiSalle (2004; 2006), and Pooley (book manuscript) – who point out that Newton was mostly concerned with a rebuttal of Descartes's notion of motion (an issue Newton was already concerned with in his pre-*Principia* manuscript *De Gravitatione* ...; see Newton (1978)). See also Koyré (1965) and Stein (1967) on Newton and Descartes.

20 Here is Mach in reaction to Newton's two globes thought experiment: "It is scarcely necessary to remark that in the reflections here presented Newton has again acted contrary to his expressed intention only to investigate *actual facts*. No one is competent to predicate things about absolute space and absolute motion; they are pure things of thought, pure mental constructs, that cannot be produced in experience. All our principles of mechanics are, as we have shown in detail, experimental knowledge concerning the relative positions and motions of bodies. Even in the provinces in which they are now recognised as valid, they could not, and were not, admitted without previously being subjected to experimental tests. No one is warranted in extending these principles beyond the boundaries of experience. In fact, such an extension is meaningless, as no one possesses the requisite knowledge to make use of it. ... When we say that a body K alters its direction and velocity solely through the influence of another body K', we have asserted a conception that it is impossible to come at unless other bodies $A, B, C...$. are present with reference to which the motion of the body K has been estimated. In reality, therefore, we are simply cognisant of a relation of the body K to $A, B, C...$. If now we suddenly neglect $A, B, C...$. and attempt to speak of the deportment of the body K in absolute space, we implicate ourselves in a twofold error. In the first place, we cannot know how K would act in the absence of $A, B, C...$.; and in the second place, every means would be wanting of forming a judgment of the behaviour of K and of putting to the test what we had predicated, – which latter therefore would be bereft of all scientific significance" (Mach, 1919, pp. 229–230).

21 Maudlin (1993) dubs the ontologically poor version of relationalism, that (in addition to temporal durations) only considers simultaneous distance relations, *Leibnizian relationalism*, and the ontologically richer alternative, that includes non-simultaneous distance relations, *Newtonian relationalism* – both should not be confounded with genuine historical positions. Note that Huggett (2006) proposes an interesting defence of Leibnizian relationalism.

22 For the historical Leibniz, this verdict is based on two fundamental metaphysical principles: the principle of sufficient reason and the principle of the identity of indiscernibles.

23 Historically, Leibniz arrived at these arguments "in a classic modus ponens/modus tollens reversal" of an argument advanced originally by Clarke against relationalism (Maudlin, 1993, p. 188).

24 Analogously, the argument can be formulated for shifts in time.

25 Leibniz himself arguably thought that the principle of the identity of indiscernibles *follows* from the principle of sufficient reason (see Section 21 in Leibniz's fifth letter to Clarke (Leibniz & Clarke, 2000, p. 53); see also Forrest,

2016; Rodriguez-Pereyra, 1999). However, Leibniz's derivation is received as unsound (Rodriguez-Pereyra, 1999). In fact, alternatively, the principle of the identity of indiscernibles is sometimes taken to be intimately related to Leibniz's thesis that concepts are complete or "individual" (see Rodriguez-Pereyra, 2014, Chapter 4; Friebe, 2022).

26 For the full argument, I refer the reader to Maudlin (1993, pp. 189–192). In short, it drives home the idea that whilst the absolute position of my chair in the actual universe *is* right here (and not 161 meters southwest), its absolute velocity in the actual universe is indeterminable.

27 See Maudlin (1990), Mundy (1992), Brighouse (1994), Rynasiewicz (1994), Hoefer (1996), Bartels (1996), and Pooley (2013).

28 In a series of publications (see Earman, 1986; 1989; Earman & Norton, 1987; Norton, 1987; 1988), John Earman and John D. Norton developed the modern version of the *hole argument* – based on what Einstein proposed in 1913 – against (certain "naïve" forms of) spacetime substantivalism in the context of general relativity. For an introduction see Norton (2019). The argument uses the diffeomorphism invariance of general relativity. Roughly speaking, it is argued that so-called *manifold substantivalism* (which identifies spacetime with the manifold of spacetime points) is committed to view spacetimes that are related by an arbitrary diffeomorphism as ontologically distinct, which leads to accepting indeterminism for general relativity. For an opposing view see Weatherall (2018) and Halvorson & Manchak (2022).

29 See, for example, Sklar (1976). See also Maudlin (1993) and Pooley (2013).

30 We usually want the symmetries of the matter field dynamics to match the symmetries of the spacetime and *vice versa* because otherwise problems of either undetectable spacetime structure or detectable but non-existent spacetime structure would arise (Earman, 1989, p. 46).

31 See, for example, Belot (2011), Pooley (2013), Martens (2019), and Menon (2019).

32 North points out that this worry is voiced (with respect to the traditional understanding) by Malament, for example: "Both positions as they are usually characterized … are terribly obscure. After they are qualified so as to seem intelligible and not too implausible, it is hard to retain a firm grasp on what divides them" (Malament, 1976, p. 317; as cited in North, 2017, pp. 128–129). Similarly, Pooley comments that "[i]t is hard to resist the suspicion that this corner of the debate is becoming merely terminological" (Pooley, 2013, p. 578).

33 This contradicts Belot (2011) who denies that the dispute between substantivalists and relationalists has *anything* to do with ontology. In particular, he does not seem to distinguish between ontology and fundamental ontology: he rejects to "simply take the characteristic difference between substantivalists and relationalists to be that the former but not the latter include space in their ontology (*or in their fundamental ontology*)" (Belot, 2011, p. 2; emphasis added). I agree that the debate is not on whether spacetime is part of the ontology (i.e., on whether spacetime exists). I disagree that the debate is neither on whether spacetime is part of the *fundamental* ontology (i.e., on whether spacetime exists fundamentally).

34 See, for example, Knox (2014; 2019), Lam and Wüthrich (2018; 2021), Baker (2021), Read (2018), and Yates (2020). To my knowledge, the term "spacetime functionalism" was first introduced by Eleanor Knox in 2014.

35 Recall that already the substantivalist defence against the shift arguments was about specifying (and arguably adjusting) what spatiotemporal aspects have to be assumed.

36 See Martens and Lehmkuhl (2020a; 2020b). See also Jaksland and Salimkhani (2023).
37 For example, a given topological structure will to some extent (mathematically) constrain what metric structure can be added.
38 See also Huggett and Wüthrich (2013), Carlip (2014), Matsubara (2017), and Wüthrich (2018).
39 The concept of emergence is highly debated in philosophy. It was originally introduced as explicitly non-reductivist; traditional accounts are due to Broad (1925), for example. Modern approaches, especially in the context of philosophy of physics, like Butterfield (2011a; 2011b) and Crowther (2013; 2015; 2016; 2018) differ drastically from this, because they typically do *not* oppose reduction but construe emergence as compatible with it – thereby capturing the generally looser use of the term in physics.
40 In metaphysical terms, for example Barnes (2012) can be viewed to accommodate a "reductivist" notion of emergence. Barnes proposes to analyse the emergent as that which is *fundamental* – it is crucial to note that she uses either the primitivist or the determination/completeness account – and *ontologically dependent* (i.e., the emergent does not exist on its own, but depends for its existence on other entities); the former captures the emergent's autonomy, the latter its reducibility. Thereby, it meets the criteria that McKenzie (2018) outlines: "The concept of metaphysical emergence is intimately tied up with our concept of fundamentality. Whether it is unpredictability, irreducibility, or metaphysical or dynamical autonomy that are taken as its hallmarks, it seems that that which characterizes the metaphysically emergent could equally characterize the fundamental. But the idea of the emergent as something arising out of complexity suggests that the concept involves the non-fundamental just as essentially. In order to get understand emergence, then, it seems we need to get a grip on what we mean by fundamentality, and how it is that we should understand the relation between the fundamental and that with which it is contrasted" (McKenzie, 2018, p. 54). Notably, Barnes's proposal is incompatible with how I conceive of fundamentality following Bennett (2017).

References

Baker, D. J. (2021). Knox's inertial spacetime functionalism (and a better alternative). *Synthese*, *199*(Suppl 2), 277–298. doi:10.1007/s11229-020-02598-z

Barnes, E. (2012). Emergence and fundamentality. *Mind*, *121*(484), 873–901. doi:10.1093/mind/fzt001

Bartels, A. (1996). Modern essentialism and the problem of individuation of spacetime points. *Erkenntnis*, *45*, 25–43. doi:10.1007/BF00226369

Bartels, A., & Wohlfarth, D. (2014). The Directedness of Time in Classical Cosmology. In G. Guo, & C. Liu (Eds.), *Scientific Explanation and Methodology of Science* (pp. 1–15). Singapore: World Scientific.

Belot, G. (2011). *Geometric Possibility*. Oxford: Oxford University Press.

Bennett, K. (2017). *Making Things Up*. Oxford: Oxford University Press.

Bricker, P. (2016). Ontological Commitment. In E. N. Zalta (Ed.), *The Stanford Encyclopedia of Philosophy* (Winter 2016 ed.). Metaphysics Research Lab, Stanford University. https://plato.stanford.edu/archives/win2016/entries/ontological-commitment/

Brighouse, C. (1994). Spacetime and holes. *PSA: Proceedings of the Biennial Meeting of the Philosophy of Science Association, 1994(1)*, 117–125. doi:10.1086/psaprocbienmeetp.1994.1.193017

Broad, C. D. (1925). *The Mind and Its Place in Nature*. London: Routledge & Kegan Paul.

Brown, H. R. (2005). *Physical Relativity: Space-Time Structure from a Dynamical Perspective*. Oxford: Oxford University Press.

Brown, H. R., & Pooley, O. (2006). Minkowski Space-Time: A Glorious Non-Entity. In D. Dieks (Ed.), *The Ontology of Spacetime. Philosophy and Foundations of Physics, Vol. 1* (pp. 67–89). Amsterdam: Elsevier.

Butterfield, J. (2011a). Emergence, reduction and supervenience: A varied landscape. *Foundations of Physics, 41(6)*, 920–959. doi:10.1007/s10701-011-9549-0

Butterfield, J. (2011b). Less is different: Emergence and reduction reconciled. *Foundations of Physics, 41(6)*, 1065–1135. doi:10.1007/s10701-010-9516-1

Carlip, S. (2014). Challenges for emergent gravity. *Studies in History and Philosophy of Science Part B: Studies in History and Philosophy of Modern Physics, 46*, 200–208. doi:10.1016/j.shpsb.2012.11.002

Carroll, S. (2004). *Spacetime and Geometry: An Introduction to General Relativity*. San Francisco, CA: Addison Wesley.

Cartwright, N. (1999). *The Dappled World*. Cambridge: Cambridge University Press.

Crowther, K. (2013). Emergent spacetime according to effective field theory: From top-down and bottom-up. *Studies in History and Philosophy of Science Part B: Studies in History and Philosophy of Modern Physics, 44(3)*, 321–328. doi:10.1016/j.shpsb.2012.08.001

Crowther, K. (2015). Decoupling emergence and reduction in physics. *European Journal for Philosophy of Science, 5(3)*, 419–445. doi:10.1007/s13194-015-0119-8

Crowther, K. (2016). *Effective Spacetime. Understanding Emergence in Effective Field Theory and Quantum Gravity*. Cham: Springer.

Crowther, K. (2018). Inter-theory relations in quantum gravity: Correspondence, reduction, and emergence. *Studies in History and Philosophy of Science Part B: Studies in History and Philosophy of Modern Physics, 63*, 74–85. doi:10.1016/j.shpsb.2017.12.002

Crowther, K., Linnemann, N., & Wüthrich, C. (2021). Spacetime functionalism in general relativity and quantum gravity. *Synthese, 199*(Suppl 2), 221–227. doi:10.1007/s11229-020-02722-z

DiSalle, R. (2004). Newton's Philosophical Analysis of Space and Time. In I. B. Cohen, & G. E. Smith (Eds.), *The Cambridge Companion to Newton* (pp. 33–56). Cambridge: Cambridge University Press.

DiSalle, R. (2006). *Understanding Space-Time*. Cambridge: Cambridge University Press.

Earman, J. (1986). Why space is not a substance (at least not to first degree). *Pacific Philosophical Quarterly, 67(4)*, 225–244. doi:10.1111/j.1468-0114.1986.tb00275.x

Earman, J. (1989). *World Enough and Space-Time*. Cambridge, MA: MIT Press.

Earman, J., & Norton, J. D. (1987). What price spacetime substantivalism? The hole story. *The British Journal for the Philosophy of Science, 38(4)*, 515–525. doi:10.1093/bjps/38.4.515

Fine, K. (2001). The question of realism. *Philosophers' Imprint, 1(1)*, 1–30. doi:2027/spo.3521354.0001.002

Forrest, P. (2016). The Identity of Indiscernibles. In E. N. Zalta (Ed.), *The Stanford Encyclopedia of Philosophy* (Winter 2016 ed.). Metaphysics Research Lab, Stanford University. https://plato.stanford.edu/archives/win2016/entries/identity-indiscernible/

Friebe, C. (2022). Leibniz, Kant, and referencing in the quantum domain. *Journal for General Philosophy of Science, 53*, 275–290. doi:10.1007/s10838-020-09515-5

Halvorson, H. & Manchak, J.B. (2022): Closing the hole argument. *The British Journal for the Philosophy of Science.* doi:10.1086/719193

Halzen, F., & Martin, A. D. (1984). *Quarks and Leptons: An Introductory Course in Modern Particle Physics*. New York, NY: John Wiley & Sons.

Hoefer, C. (1996). The metaphysics of space-time substantivalism. *The Journal of Philosophy, 93(1)*, 5–27. doi:10.2307/2941016

Huggett, N. (2006). The regularity account of relational spacetime. *Mind, 115(457)*, 41–73. doi:10.1093/mind/fzl041

Huggett, N. (2017). Target space ≠ space. *Studies in History and Philosophy of Science Part B: Studies in History and Philosophy of Modern Physics, 59*, 81–88. doi:10.1016/j.shpsb.2015.08.007

Huggett, N., & Wüthrich, C. (2013). Emergent spacetime and empirical (in) coherence. *Studies in History and Philosophy of Science Part B: Studies in History and Philosophy of Modern Physics, 44(3)*, 276–285. doi:10.1016/j.shpsb.2012.11.003

Jaksland, R., & Salimkhani, K. (2023). The many problems of spacetime emergence in quantum gravity, *The British Journal for the Philosophy of Science.*

Kiefer, C. (2007). *Quantum Gravity*. Oxford: Oxford University Press.

Kim, J. (1998). *Mind in a Physical World: An Essay on the Mind-Body Problem and Mental Causation*. Cambridge, MA: MIT Press.

Knox, E. (2013). Effective spacetime geometry. *Studies in History and Philosophy of Science Part B: Studies in History and Philosophy of Modern Physics, 44*, 346–356. doi:10.1016/j.shpsb.2013.04.002

Knox, E. (2014). Newtonian spacetime structure in light of the equivalence principle. *The British Journal for the Philosophy of Science, 65(4)*, 863–880. doi:10.1093/bjps/axt037

Knox, E. (2019). Physical relativity from a functionalist perspective. *Studies in History and Philosophy of Science Part B: Studies in History and Philosophy of Modern Physics, 67*, 118–124. doi:10.1016/j.shpsb.2017.09.008

Koyré, A. (1965). *Newtonian Studies*. Chicago, IL: University of Chicago Press.

Lam, V., & Wüthrich, C. (2018). Spacetime is as spacetime does. *Studies in History and Philosophy of Science Part B: Studies in History and Philosophy of Modern Physics, 64*, 39–51. doi:10.1016/j.shpsb.2018.04.003

Lam, V., & Wüthrich, C. (2021). Spacetime functionalism from a realist perspective. *Synthese, 199*(Suppl 2), 335–353. doi:10.1007/s11229-020-02642-y

Laymon, R. (1978). Newton's bucket experiment. *Journal of the History of Philosophy, 16(4)*, 399–413. doi:10.1353/hph.2008.0681

Le Bihan, B., & Linnemann, N. (2019). Have we lost spacetime on the way? Narrowing the gap between general relativity and quantum gravity. *Studies in*

History and Philosophy of Science Part B: Studies in History and Philosophy of Modern Physics, 65, 112–121. doi:10.1016/j.shpsb.2018.10.010

Lehmkuhl, D. (2018). The metaphysics of super-substantivalism. *Noûs*, *52(1)*, 24–46. doi:10.1111/nous.12163

Leibniz, G. W., & Clarke, S. (2000). *Correspondence* (R. Ariew, Ed.). Indianapolis, IN/Cambridge: Hackett Publishing Company, Inc.

Linnemann, N. (2021). On the empirical coherence and the spatiotemporal gap problem in quantum gravity: And why functionalism does not (have to) help. *Synthese*, *199*(Suppl 2), 395–412. doi:10.1007/s11229-020-02659-3

Mach, E. (1919). *The Science of Mechanics: A critical and Historical Account of Its Development* (4th ed.). Chicago, IL/London: The Open Court Publishing Co.

Malament, D. B. (1976). Review of Lawrence Sklar, space, time, and spacetime. *Journal of Philosophy*, *73(11)*, 306–323. doi:10.2307/2025892

Martens, N. C. (2019). The metaphysics of emergent spacetime theories. *Philosophy Compass*, *14(7)*, e12596. doi:10.1111/phc3.12596

Martens, N. C., & Lehmkuhl, D. (2020a). Cartography of the space of theories: An interpretational chart for fields that are both (dark) matter and spacetime. *Studies in History and Philosophy of Science Part B: Studies in History and Philosophy of Modern Physics*, *72*, 217–236. doi:10.1016/j.shpsb.2020.08.004

Martens, N. C., & Lehmkuhl, D. (2020b). Dark matter = modified gravity? Scrutinising the spacetime–matter distinction through the modified gravity/dark matter lens. *Studies in History and Philosophy of Science Part B: Studies in History and Philosophy of Modern Physics*, *72*, 237–250. doi:10.1016/j.shpsb.2020.08.003

Matsubara, K. (2017). Quantum gravity and the nature of space and time. *Philosophy Compass*, *12(3)*, e12405. doi:10.1111/phc3.12405

Maudlin, T. (1990). Substances and space-time: What Aristotle would have said to Einstein. *Studies in History and Philosophy of Science Part A*, *21(4)*, 531–561. doi:10.1016/0039-3681(90)90032-4

Maudlin, T. (1993). Buckets of water and waves of space: Why spacetime is probably a substance. *Philosophy of Science*, *60(2)*, 183–203. doi:10.1086/289728

Maudlin, T. (2007). *The Metaphysics Within Physics*. Oxford: Oxford University Press.

McKenzie, K. (2011). Arguing against fundamentality. *Studies in History and Philosophy of Science Part B: Studies in History and Philosophy of Modern Physics*, *42(4)*, 244–255. doi:10.1016/j.shpsb.2011.09.002

McKenzie, K. (2017). Against Brute fundamentalism. *Dialectica*, *71*, 231–261. doi:10.1111/1746-8361.12189

McKenzie, K. (2018). Fundamentality. In S. Gibb, R. F. Hendry, & T. Lancaster (Eds.), *The Routledge Handbook of Emergence* (pp. 54–64). London/New York, NY: Routledge.

McKenzie, K. (2022). *Fundamentality and Grounding*. Cambridge: Cambridge University Press.

McLaughlin, B., & Bennett, K. (2018). Supervenience. In E. N. Zalta (Ed.), *The Stanford Encyclopedia of Philosophy* (Winter 2018 ed.). Metaphysics Research Lab, Stanford University. https://plato.stanford.edu/archives/win2018/entries/supervenience/

Menon, T. (2019). Algebraic fields and the dynamical approach to physical geometry. *Philosophy of Science, 86(5)*, 1273–1283. doi:10.1086/705508

Morrison, M. (2006). Emergence, reduction, and theoretical principles: Rethinking fundamentalism. *Philosophy of Science, 73(5)*, 876–887. doi:10.1086/518746

Mundy, B. (1992). Space-time and isomorphism. *PSA: Proceedings of the Biennial Meeting of the Philosophy of Science Association, 1992*, 515–527. doi:10.1086/psaprocbienmeetp.1992.1.192780

Nagel, E. (1961). *The Structure of Science*. New York, NY: Harcourt, Brace & World, Inc.

Nerlich, G. (1994). *The Shape of Space* (2nd ed.). Cambridge: Cambridge University Press.

Newton, I. (1846). *The Mathematical Principles of Natural Philosophy*. New York, NY: Daniel Adee.

Newton, I. (1978). De Gravitatione et Aequipondio Fluidorum. In A. R. Hall, & M. B. Hall (Eds.), *Unpublished Scientific Papers of Isaac Newton* (First paperback ed., pp. 89–156). Cambridge: Cambridge University Press.

Ney, A. (2008). Reductionism. In J. Fieser, & B. Dowden (Eds.), *The Internet Encyclopedia of Philosophy*. https://www.iep.utm.edu/red-ism/

North, J. (2018). A New Approach to the Relational–Substantival Debate. In K. Bennett, & D. W. Zimmerman (Eds.), *Oxford Studies in Metaphysics, Vol. II* (pp. 3–43). Oxford: Oxford University Press.

Norton, J. D. (1987). Einstein, The Hole Argument and the Reality of Space. In J. Forge (Ed.), *Measurement, Realism and Objectivity: Essays on Measurement in the Social and Physical Sciences. Australasian Studies in History and Philosophy of Science, Vol. 5* (pp. 153–188). Dordrecht: Springer.

Norton, J. D. (1988). The hole argument. *PSA: Proceedings of the Biennial Meeting of the Philosophy of Science Association, 1988(2)*, 56–64. doi:10.1086/psaprocbienmeetp.1988.2.192871

Norton, J. D. (2008). Why constructive relativity fails. *The British Journal for the Philosophy of Science, 59(4)*, 821–834. doi:10.1093/bjps/axn046

Norton, J. D. (2019). The Hole Argument. In E. N. Zalta (Ed.), *The Stanford Encyclopedia of Philosophy* (Summer 2019 ed.). Metaphysics Research Lab, Stanford University. https://plato.stanford.edu/archives/sum2019/entries/spacetime-holearg/

Paul, L. A. (2012). Building the world from its fundamental constituents. *Philosophical Studies, 158*, 221–256. doi:10.1007/s11098-012-9885-8

Pooley, O. (2013). Substantivalist and Relationalist Approaches to Spacetime. In R. Batterman (Ed.), *The Oxford Handbook of Philosophy of Physics*. Oxford: Oxford University Press.

Pooley, O. (Book manuscript). *The reality of spacetime*.

Quine, W. V. (1951). Ontology and ideology. *Philosophical Studies, 2(1)*, 11–15. doi:10.1007/BF02198233

Read, J. (2018). Functional gravitational energy. *The British Journal for the Philosophy of Science, 71(1)*, 205–232. doi:10.1093/bjps/axx048

Read, J. (2019). On miracles and spacetime. *Studies in History and Philosophy of Science Part B: Studies in History and Philosophy of Modern Physics, 65*, 103–111. doi:10.1016/j.shpsb.2018.10.002

Read, J. (2020). Explanation, Geometry, and Conspiracy in Relativity Theory. In C. Beisbart, T. Sauer, & C. Wüthrich (Eds.), *Thinking About Space and Time: 100 Years of Applying and Interpreting General Relativity. Einstein Studies, Vol. 15* (pp. 173–205). Basel: Birkhäuser.

Read, J., Brown, H. R., & Lehmkuhl, D. (2018). Two miracles of general relativity. *Studies in History and Philosophy of Science Part B: Studies in History and Philosophy of Modern Physics, 64*, 14–25. doi:10.1016/j.shpsb.2018.03.001

Rodriguez-Pereyra, G. (1999). Leibniz's argument for the identity of indiscernibles in his correspondence with Clarke. *Australasian Journal of Philosophy, 77(4)*, 429–438. doi:10.1080/00048409912349201

Rodriguez-Pereyra, G. (2014). *Leibniz's Principle of Identity of Indiscernibles.* Oxford: Oxford University Press.

Rolffs, M. (2023). *Kausalität und mentale Verursachung. Eine Verteidigung des nicht-reduktiven Physikalismus.* Berlin, Heidelberg: J.B. Metzler. doi:10.1007/978-3-662-66778-1

Rynasiewicz, R. (1994). The lessons of the hole argument. *The British Journal for the Philosophy of Science, 45(2)*, 407–436. doi:10.1093/bjps/45.2.407

Rynasiewicz, R. (1995a). By their properties, causes and effects: Newton's Scholium on time, space, place and motion—I. The text. *Studies in History and Philosophy of Science Part A, 26(1)*, 133–153. doi:10.1016/0039-3681(94)00035-8

Rynasiewicz, R. (1995b). By their properties, causes and effects: Newton's Scholium on time, space, place and motion—II. The context. *Studies in History and Philosophy of Science Part A, 26(2)*, 295–321. doi:10.1016/0039-3681(94)00049-F

Rynasiewicz, R. (2014). Newton's Views on Space, Time, and Motion. In E. N. Zalta (Ed.), *The Stanford Encyclopedia of Philosophy* (Summer 2014 ed.). Metaphysics Research Lab, Stanford University. https://plato.stanford.edu/archives/sum2014/entries/newton-stm/

Schaffer, J. (2009). On What Grounds What. In D. Manley, D. J. Chalmers, & R. Wasserman (Eds.), *Metametaphysics: New Essays on the Foundations of Ontology* (pp. 347–383). Oxford: Oxford University Press.

Sider, T. (2011). *Writing the Book of the World.* Oxford: Clarendon Press.

Sklar, L. (1976). *Space, Time, and Spacetime.* Berkeley and Los Angeles, CA: University of California Press.

Stein, H. (1967). Newtonian space-time. *Texas Quarterly, 10*, 174–200.

Thomson, M. (2013). *Modern Particle Physics.* Cambridge: Cambridge University Press.

van Riel, R., & Van Gulick, R. (2019). Scientific Reduction. In E. N. Zalta (Ed.), *The Stanford Encyclopedia of Philosophy* (Spring 2019 ed.). Metaphysics Research Lab, Stanford University. https://plato.stanford.edu/archives/spr2019/entries/scientific-reduction/

Wald, R. M. (1984). *General Relativity.* Chicago, IL: The University of Chicago Press.

Weatherall, J. O. (2018). Regarding the 'hole argument'. *The British Journal for the Philosophy of Science, 69(2)*, 329–350. doi:10.1093/bjps/axw012

Weinberg, S. (1977). The forces of nature. *American Scientist, 65*, 171–176.

Weinberg, S. (1993). *Dreams of a Final Theory.* London: Hutchinson Radius.

Weinberg, S., & Witten, E. (1980). Limits on massless particles. *Physics Letters B, 96(1–2)*, 59–62. doi:10.1016/0370-2693(80)90212-9

Wilson, J. M. (2014). No work for a theory of grounding. *Inquiry, 57(5–6)*, 535–579. doi:10.1080/0020174X.2014.907542

Wüthrich, C. (2018). The Emergence of Space and Time. In S. Gibb, R. F. Hendry, & T. Lancaster (Eds.), *The Routledge Handbook of Emergence* (pp. 315–326). London/New York: Routledge.

Yates, D. (2020). Thinking About Spacetime. In C. Wüthrich, B. L. Bihan, & N. Huggett (Eds.), *Philosophy Beyond Spacetime: Implications from Quantum Gravity* (pp. 129–153). Oxford: Oxford University Press. doi:10.1093/oso/9780198844143.003.0006

3 Spacetime in Relativity Theory

3.1 Brief Recap of Relativity Theory

3.1.1 General Relativity

Einstein's theory of general relativity is our best theory of gravitation and spacetime.[1] Mathematically it involves the study of a four-dimensional pseudo-Riemannian manifold, which is a differentiable manifold M that is equipped with a metric tensor $g_{\mu\nu}$ that is symmetric ($g_{\mu\nu} = g_{\nu\mu}$) and non-singular ($\det g_{\mu\nu} > 0$) at every point of the manifold.[2] According to the standard interpretation, spacetime is then taken to be represented by the pair $\langle M, g \rangle$ of manifold structure M and metric structure g.[3] As it is instructive, let me briefly (and roughly) lay out how one gets from less organised mathematical structures to pseudo-Riemannian manifolds following Carroll (2004) and Norton (2019).

The empirical basis of general relativity is point-like *events*. Naturally, the least organised structure is a *set* of point-like events (or, a set of points associated with the events). But there is more to describing our universe than just listing which and how many events there are. Ultimately, we want to say things like: event A, say, a burst of neutrinos detected on Earth, occurred three hours before event B, say, the detection by Earth-bound observatories of the first visible light from the collapse of supernova SN1987A in the Large Magellanic Cloud, 168,000 light-years distant from Earth. But before we get to measure spatiotemporal distances, we first need to promote the set of point-like events (or, rather the set of points associated with the events) to a basic notion of a space and, subsequently, define a metric on that space. This is done by first introducing the fundamental notion of an *open set* which defines a *topological space* from the set of events (or rather from the set of points associated with the events). An open set is a subset of a set of points in which every element of the subset is only surrounded by other elements of the subset, which is why the boundary is not part of the subset. In other words, the open set structure

DOI: 10.4324/9781003404149-3

introduces the notion of a *neighbourhood*. For describing the universe, we now need to appropriately "sew together" many of these open sets to yield what might be called a "neighbourhood of neighbourhoods"; with respect to events in our universe, we also have the idea that some events are local neighbours, say, events on Earth, which are situated themselves in a larger neighbourhood, say, our solar system, and so on. By additionally demanding that each open set "looks like" $\mathbb{R}^n$ (with n being fixed for all open sets) and demanding that the sewing procedure is smooth, we obtain a *manifold* as the full topological space. The neighbourhood relations between events are expressed by the fact that we can smoothly label the events with coordinates. The number of coordinates needed for this, here four, indicates the dimensionality of the manifold. But only by adding (Lorentzian) metrical structure to the topological manifold structure do we arrive at an (appropriate) notion of *spatiotemporal distance*, namely $\Delta s = x^\mu x_\mu = g_{\mu\nu} x^\mu x^\nu$. If one only demands that the metric be non-singular everywhere on the manifold, this space is called a pseudo-Riemannian manifold.

But we are not quite done yet. We still need to introduce one additional structure (which is needed for anything that involves objects from different tangent spaces of the manifold – taking covariant derivatives, for example): the *(affine) connection*. On curved manifolds, a connection is essential for comparing vectors and tensors from different tangent spaces, i.e., at different points of the manifold. A connection defines a notion of "keeping a tensor constant along some path", i.e., a notion of "parallel transporting" a tensor. It is via the notion of parallel transport defined by the connection that we have a reference for evaluating, for example, the rate of change of a tensor field along some path (given by its covariant derivative). A pseudo-Riemannian manifold automatically yields a connection that is derived directly from the metric: the unique torsion-free ($\Gamma^\lambda_{\mu\nu} = \Gamma^\lambda_{(\mu\nu)}$) and metric compatible ($D_\rho g_{\mu\nu} = 0$, with the covariant derivative D_μ) Levi-Civita connection Γ with Christoffel symbols

$$\Gamma^\lambda_{\mu\nu} = \frac{1}{2} g^{\lambda\sigma} \left(\partial_\mu g_{\nu\sigma} + \partial_\nu g_{\sigma\mu} - \partial_\sigma g_{\mu\nu} \right) \tag{3.1}$$

The affine connection is required to be metric compatible, because only then are geodesics in the affine sense also geodesics in the metric sense; "the worldlines of freely falling bodies are geodesics of the connection" (Brown, 2005, p. 118).

Note that one could instead have defined the connection independently of the metric. In fact, one can in principle define several connections – and metrics – on any manifold, which is important for so-called bimetric theories of gravity (see Rosen, 1940a, 1940b).[4] Treating the connection as an object independent of the metric is prominently used in the so-called

Palatini formalism which derives the Einstein field equations by varying the Palatini action with respect to g *and* Γ as independent variables.[5]

Standardly, the Einstein field equations are derived from the Einstein-Hilbert action; a third derivation is sketched in Chapter 4. The Einstein field equations (without cosmological constant term) are

$$R_{\mu\nu} - \frac{1}{2} R \, g_{\mu\nu} = 8\pi G \, T_{\mu\nu} \tag{3.2}$$

with the Riemann curvature tensor

$$R^{\rho}_{\sigma\mu\nu} = \partial_\mu \Gamma^{\rho}_{\nu\sigma} - \partial_\nu \Gamma^{\rho}_{\mu\sigma} + \Gamma^{\rho}_{\mu\lambda}\Gamma^{\lambda}_{\nu\sigma} - \Gamma^{\rho}_{\nu\lambda}\Gamma^{\lambda}_{\mu\sigma} \tag{3.3}$$

the Ricci tensor $R_{\mu\nu} = R^{\alpha}_{\alpha\mu\nu}$, the Ricci scalar $R = R^{\mu}_{\mu}$, the gravitational constant G, and the energy-momentum tensor $T_{\mu\nu}$. The left-hand side is often written as the Einstein tensor $G_{\mu\nu} = R_{\mu\nu} - \frac{1}{2} R \, g_{\mu\nu}$. For a given energy-momentum tensor, the Einstein field equations are solved by a generally curved metric field $g_{\mu\nu}$. A spacetime is curved if not all components of the Riemann curvature tensor vanish; spacetime is considered to be curved in general relativity (as opposed to flat Euclidean space in Newtonian mechanics or flat Minkowski spacetime in special relativity, for example).

I shall discuss the standard textbook interpretation in more detail in a moment, but note already here that the Einstein field equations provide us with a first glance at why general relativity is typically taken to "geometrise" gravity: the left-hand side of Eq. (3.2) features objects that seem to have a natural interpretation in terms of spacetime geometry – most importantly, the metric field g representing (metrical aspects of) spacetime; the source term on the right-hand side, which features the energy-momentum tensor, contains the matter content. Therefore, the Einstein field equations suggest that matter and energy determine the – generally curved – geometry of spacetime.[6] In this curved spacetime, free massive point particles can be shown to move along so-called *geodesics*[7] defined by the geometry of spacetime.[8] According to the standard interpretation of general relativity, a planet's stably orbiting its star is not due to two balanced forces, but due to the fact that the planet follows its geodesic motion *freely* – gravity seems to be a "pseudo force".

Since I will be referring to Einstein's theory of special relativity throughout this chapter, partly as a helpful toy model for better understanding the different interpretations of general relativity, let me also quickly summarise some of its central aspects. In fact, in the course of this work, I shall propose an understanding of general relativity as essentially of the special-relativistic type (see Chapter 4) – in this scenario, special relativity obviously is not a toy model anymore.

3.1.2   Special Relativity

Conceptually, Einstein (1905)[9] based special relativity on the following two postulates: the principle of relativity, stating that the laws of physics are the same in all inertial frames,[10] and the light postulate, stating that in empty space light propagates with a constant velocity c regardless of the state of motion of the emitting source. In a modern reading, this is equivalent to demanding *Lorentz invariance* (the invariance under global Lorentz transformations) for all physical processes.

What is important is that, due to Hermann Minkowski,[11] there is a spatiotemporal interpretation of special relativity, which understands special relativity as the theory of four-dimensional *Minkowski spacetime* – a flat manifold M equipped with a flat, symmetric, and non-singular metric $\eta_{\mu\nu}$. It is "flat" in the sense that all components of the Riemann curvature tensor vanish at every point of the manifold. Essentially, general-relativistic spacetime is locally approximated by Minkowski spacetime – the global inertial frames of special relativity become local inertial frames in general relativity.[12]

Since the speed of light c is constant, Minkowski spacetime (and general-relativistic spacetime as well) is equipped with a *light cone structure* that distinguishes null or lightlike, timelike, and spacelike curves. Lightlike curves lie on the boundary of the light cone and represent trajectories for objects moving at the speed of light. All lightlike curves through a spacetime point constitute the past-directed and the future-directed light cone at that spacetime point. Timelike curves stay within the light cone and represent trajectories for objects moving at velocities below the speed of light. Finally, spacelike curves are all curves outside the light cone and do not represent physical trajectories – at most hypothetically for objects travelling faster than the speed of light.

3.2   Spacetime Fundamentalism in Relativity Theory

Now, what do special and general relativity say with respect to the issue of whether spacetime is fundamental? As I have noted earlier, spacetime fundamentalism is the mainstream position with respect to all major theories of spacetime, including Newtonian theory and special and general relativity. With the possible exception of general relativity, this is not so much because relationalism is inconceivable within these theories, but because it is, at least in part, a less natural way of making sense of them. To pick up on my discussion in Section 2.2.1, let me now briefly highlight in which sense this is true for relativity theory, and how both theories, nevertheless, mark important shifts in the traditional substantivalism–relationalism debate.

But before I start, let me highlight again that the standard interpretation of spacetime in special and general relativity goes by different names. In varying terminology, it is understood as the *geometrical, realist, substantivalist*, or *fundamentalist* interpretation. In this work I take it that these notions essentially refer to the same position, although they might stress different aspects of it. While the term "geometrical view" is generally used to underline that the standard interpretation conceives of spacetime geometry as *explanatory* of gravitational effects and, more importantly, the dynamical properties of matter fields, the other terms draw attention to the *ontological* aspect, namely that spacetime is viewed as an *ontologically independent* entity – in particular, as ontologically independent of matter fields.

Standardly, special and general relativity are straightforwardly interpreted as positing spacetime as a fundamental structure. In particular, it seems natural to view the manifold M as representing spacetime. With all additional structures, e.g., metric structure and other field structure, being defined on the manifold, this view is arguably closest to the traditional "container" view of spacetime (see Norton, 2019).[13] However, as emphasised in Section 2.2.1, especially in light of the hole argument there has been a significant shift towards also including metric structure for representing spacetime. Accordingly, the standard picture identifies spacetime as the pair of manifold and metric structure, viz., $\langle M, g \rangle$ for general relativity, and $\langle M, \eta \rangle$ for special relativity.

For special relativity, Norton summarises this standard view as the "realist conception of Minkowski spacetime" (Norton, 2008, p. 823) which implies the following[14]:

(1) There is a four-dimensional spacetime that can be coordinatized by a set of standard coordinates (x, y, z, t), related by the Lorentz transformation.

(2) The spatiotemporal interval s between events (x, y, z, t) and (X, Y, Z, T) along a straight line connecting them is a property of the spacetime, independent of the matter it contains, and is given by

$$s^2 = (t - T)^2 - (x - X)^2 - (y - Y)^2 - (z - Z)^2.$$

When $s^2 > 0$, the interval s corresponds to times elapsed on an ideal clock; when $s^2 < 0$, the interval s corresponds to spatial distances measured by ideal rods (both employed in the standard way).

(3) Material clocks and rods measure these times and distances because the laws of the matter theories that govern them are adapted to the independent geometry of this spacetime.

(Norton, 2008, p. 823)

In light of Section 2.2.2, however, the notion "realist interpretation" is arguably a misnomer; it would be more accurate to refer to it as what I have previously dubbed spacetime fundamentalism. In fact, also Norton's own formulation "*independent of*" indicates that it is about fundamentality, i.e., (ontological) independence. In fact, Menon (2019) – albeit for a different reason – rephrases Norton's definition as follows (boldface emphasis is due to Menon to highlight his modifications):

1. There exists a **fundamental** four-dimensional spacetime that can be coordinatized by a set of standard coordinates (x, y, z, t) related by the Lorentz transformation.
2. The spatiotemporal interval s between events (x, y, z, t) and (X, Y, Z, T) along a straight line connecting them is a property of the spacetime, **ontologically** independent of the matter it contains, and is given by $s^2 = (t - T)^2 - (x - X)^2 - (y - Y)^2 - (z - Z)^2$.
3. Material clocks and rods measure these times and distances because the laws of the matter theories that govern them are adapted to the **ontologically** independent geometry of this spacetime.
                                                                    (Menon, 2019, p. 1278)

This can be translated into the following general characterisation of Nortonian spacetime fundamentalism in relativity theory (notice endnote 14), which, in particular, also applies to general relativity: the spacetime fundamentalist interprets special and general relativity as being committed to spacetime as the pair of

(A) a fundamental, i.e., ontologically independent, manifold (representing topological properties), and
(B) a fundamental, i.e., ontologically independent, metric field (representing metrical properties).
(C) Material rods and clocks built from the matter fields survey the metric field, because the symmetry properties of the matter field dynamics coincide with (or are adapted to) the ontologically independent symmetry properties of the metric field.[15]

Regarding (C), I have explicated Norton's notion of "adaptation" in terms of coinciding symmetry properties, which is in accordance with the terminology used in the dynamical approach literature (see Section 3.5). Naturally, this prompts the request for an explanation: how is it that the respective symmetry properties coincide? Although one might read the spacetime fundamentalist as remaining neutral on this point, I take it that the fundamentalist will typically prefer to view the symmetry properties of spacetime (or, more precisely, the metric) as *explanatorily prior* to the symmetry properties of the matter field dynamics; this is often dubbed the *geometrical view*. In fact, Norton agrees that the fundamentalist (or, in his words, the realist) does not opt for a "mutual adaption" (Norton, 2008, p. 830) between matter and spacetime, but "specifies a direction of adaption: ... matter is adapted to spacetime" (Norton, 2008, p. 830) – meaning that the spacetime symmetries are prior. Accordingly, (C) contains the ontological claim, that the metric field has the *intrinsic* property of *being the metric that is surveyed by material rods and clocks*; this property is dubbed the metric's *chronogeometricity* (for more details see Section 3.5). Note that a field's having metrical properties is not the same as a field's having chronogeometricity: having metrical properties only means that the field has the general property of being able to yield spatiotemporal distances (technically: of being represented by a metric tensor). In other words, having metrical properties means to be *a* metric field, while having also chronogeometricity means to be *the* metric field which is actually surveyed by material rods and clocks. (It may be helpful to recall that there can in principle be several metric fields on a manifold.)

Note that contesting the explanatory and ontological claims of (C) – as is done by the dynamical approach – is usually *not* understood as a commitment to contesting the ontological claims (A) and (B) as well. The dynamical approach to *special* relativity does oppose not only (C) but also (B): the Minkowski metric η is argued to be non-fundamental, i.e., not part of the fundamental ontology; in fact, Brown and Pooley (2006) even claim that η is a non-entity, i.e., not part of the ontology at all. On the contrary, the dynamical approach to *general* relativity opposes only (C): the g field remains fundamental on the dynamical approach, although its chronogeometricity is non-fundamental. Separating these two aspects is at the heart of the recent debate on the dynamical approach, which, in this sense, is indeed a genuinely new debate. The traditional substantivalism–relationalism debate lurks in the background, though. I shall gently push back on the prevalent view that the two debates are distinct. For a discussion of how to contest (A), see Section 3.6.

To summarise, the standard view is that spacetime (i.e., the pair of manifold structure and metric structure) is fundamental according to special and general relativity. In particular, special relativity's Minkowski spacetime is arguably closest to what one could call an "absolute" or "container-like" spacetime, since the spacetime structure is fixed. The metric field $\eta_{\mu\nu}$ is identically the same in all kinematically possible models (KPMs) (Pooley, 2017, p. 13).

That said, it is worth pointing out that it is, in fact, special relativity which alternatively admits the most natural relationalist (or constructivist) interpretation, namely in terms of what Maudlin dubs *Minkowski relationalism* (see Earman, 1989, pp. 128–130; Maudlin, 1993, pp. 196–199). In short, Minkowski relationalism posits a fundamental ontology of particles and special-relativistic spatiotemporal relations[16]; the dynamical approach to special relativity can be seen as another example of how to be a relationalist with regard to special relativity (see Pooley, 2013).[17] Contrary to the standard view, Minkowski spacetime can then be argued to be *non-fundamental*.

However, trying to repeat this "triumph" (Maudlin, 1993, p. 199) of Minkowski relationalism in the context of general relativity is typically deemed a "hopeless task" (Maudlin, 1993, p. 199). Essentially, this is because the spatiotemporal structure of general-relativistic spacetime is not fixed, but dynamical, and depends on the matter distribution:

> Ironically it is exactly the absoluteness of Newtonian space and of Minkowski spacetime that allows the relationist to treat them as fictions yet still exploit their mathematical utility. Because the geometrical structure of these spacetimes is fixed independently of the matter in them, one knows a priori the nature of the manifold into which the particle trajectories must be embedded.
>
> (Maudlin, 1993, p. 199)

In particular, it is not sufficient to fix the relations of material particles to fix the general-relativistic spacetime geometry:

> For example, no amount of information about the past history of a set of particles can determine whether a gravitational wave is approaching from outside the system, a wave whose presence has not been recorded on any of the material world-lines.
>
> (Maudlin, 1993, p. 199)

To remedy this deficiency, the relationalist may consider extending her ontology by generalising her notion of a "material particle" to make accessible information about gravitational waves, for example. Arguably, the relationalist is then ultimately forced to "adopt some form of plenism" (Maudlin, 1993, p. 200), namely in terms of a field ontology. The most natural candidate for such a field is the metric g. This, however, is spacetime fundamentalism.[18]

Moreover, spacetime fundamentalism receives critical support from the fact that non-trivial vacuum solutions of general relativity not only exist, but are also physically relevant (e.g., they allow for gravitational waves). According to Maudlin (1993), the very fact that general relativity predicts gravitational waves – we can now say: that there are gravitational waves (Abbott et al., 2016) – supports spacetime fundamentalism; Chapters 4 and 5 revisit this issue from the opposite perspective. The fact that general relativity tells us that "empty spacetime" is physically interesting does seem to bolster spacetime fundamentalism.

So making a case against fundamentalism arguably requires additional resources. For example, the non-fundamentalist may try to find explanatory gaps in the fundamentalist account, or argue that the account is problematic for reasons external to general relativity. Also, in light of the discussion above, the non-fundamentalist may try to find a fixed-field formulation of general relativity. We encounter these strategies again in Section 3.5 and Chapters 4 and 5, when considering how the dynamical approach challenges fundamentalism and how it may utilise foundational insights from other parts of physics. But first, to also briefly review the physical basics of general relativity, let me present the "textbook" reason for why spacetime is standardly taken to be fundamental in general relativity.

3.3 Geometrisation of Gravity[19]

In the usual pseudo-Riemannian formulation of general relativity, basic notions like "metric" and "curvature" – and really the whole apparatus of pseudo-Riemannian geometry – strongly suggest a view that grants the metric field g (together with the manifold M) the status of spacetime *per se*, and then explains the gravitational phenomena by the geometry of this spacetime. Test particles freely follow the geodesics of curved spacetime. Gravitational effects between the particles show "in the deviation of one geodesic from a nearby geodesic ('relative acceleration of test particles')" (Misner et al., 1973, p. 195).

According to this standard interpretation, general relativity *geometrises* gravity. Gravity is not a "force" or interaction represented by a gravitational field *in* spacetime, like Newtonian theory has it (or like electromagnetism is represented by the electromagnetic field), but is reduced to spacetime geometry itself: gravity is a mere effect of the curvature of spacetime, a geometrised pseudo force. By contrast, the standard lore about Newtonian theory is that it "casts gravity as a force. That is, gravity causes objects to deviate from inertial trajectories. Newtonian gravity is not a consequence of the geometry of space and time; forces and fields propagate in fixed Euclidean space" (Knox, 2011, p. 264). Note, however, that the issue is more complicated, since using differential geometry is not restricted to general relativity: also Newtonian gravity can be written "in such a way that Newtonian gravity, as in GR [general relativity], appears to be a manifestation of geometrical spacetime structure", which is known as *Newton–Cartan theory* (Knox, 2011, p. 264). In fact, so-called *teleparallel gravity* turns this around and reformulates general relativity as a force theory (Knox, 2011, p. 264).

The interpretation of general relativity as a reduction of gravitation to spacetime curvature is often attributed to Einstein himself (Weinberg,

1972, pp. vii, 147). Lehmkuhl (2014), however, presents new sources that do not support this view. As a matter of fact, general relativity is usually presented as a geometrisation of gravity in textbooks:

> General relativity ... is Einstein's theory of space, time, and gravitation. At heart it is a very simple subject (compared, for example, to anything involving quantum mechanics). The essential idea is perfectly straightforward: while most forces of nature are represented by fields defined on spacetime (such as the electromagnetic field, or the short-range fields characteristic of subnuclear forces), gravity is inherent in spacetime itself. In particular, what we experience as "gravity" is a manifestation of the curvature of spacetime. Our task, then, is clear. We need to understand spacetime, we need to understand curvature, and we need to understand how curvature becomes gravity.
>
> (Carroll, 2004, p. 1)

Misner et al. (1973, p. 195) interpret gravitation as "a manifestation of spacetime curvature" as well. Similarly, Wald (1984, p. 64) argues that "we are forced to view gravity as an aspect of spacetime structure" (see also Lehmkuhl, 2008). Also many philosophers advocate the standard view – for example, Friedman (1983), Maudlin (1996, 2012), Nerlich (1994), and, to a lesser extent, Norton (2019).

Undoubtedly, the standard interpretation is very appealing. First, it remains close to the standard mathematical formalism that successfully unifies two apparently very different concepts: gravitation and spacetime geometry. Second, it yields quite transparent ontological commitments: most importantly, that spacetime is a generally curved pseudo-Riemannian manifold M with a metric g and that gravitation is not due to an additional, ontologically self-standing field in spacetime (like electromagnetism is due to the electromagnetic field), but simply the result of spacetime's being curved. On the face of it, this is a perfectly fine and "deeply seductive" (Knox, 2011, p. 269) interpretation of general relativity.

But why adopt this interpretation? What underpins the geometrisation picture of gravity and the underlying interpretation of g as spacetime (or the metrical aspect thereof) except the fact that the standard formalism of general relativity makes use of pseudo-Riemannian geometry, which, on its own, is hardly convincing? As I have stressed above and shall repeat below, there are other formulations that do not wear this interpretation on their sleeve. To answer this, one needs to look at the conceptual basis of general relativity. At the core of general relativity, and at the core of the geometrisation of gravity, we find Einstein's principle of equivalence; for the moment, I only give a rough overview, but shall discuss the equivalence principle in more detail in Section 5.2.3.

Note that the equivalence principle comes in different variants. Essentially, there are weak and strong versions (e.g., Carroll, 2004, pp. 48–54). A weak version of the equivalence principle is already familiar from Newtonian mechanics and states that the inertial mass m_i and the gravitational mass m_g of any object are equal in value.[20] In Newtonian mechanics the inertial mass is the constant of proportionality between some force and the acceleration of the object the force acts on. Since the value of the inertial mass of the object is the same for any force, the inertial mass is universal in character (Carroll, 2004, p. 48). On the other hand, the gravitational mass is a specific quantity only related to the gravitational force – it is the constant of proportionality between the gradient of the gravitational potential and the gravitational force (Carroll, 2004, p. 48). *Prima facie*, both masses are conceptually independent. Hence, m_g/m_i may differ for different objects and may therefore be thought of as a "gravitational charge" (Carroll, 2004, p. 48). Accordingly, the behaviour of different objects in a gravitational field would generally depend on the (different) gravitational charges, just as the behaviour of electromagnetically charged particles in an electromagnetic field depends on the particles' charges. However, since Galileo we *empirically* know that inertial and gravitational mass are always equal in value, i.e., $m_i = m_g$.[21] Every object in a gravitational field falls at the same rate regardless of its internal properties – in particular, regardless of its mass.[22] Thus, in Newtonian mechanics inertial and gravitational mass are conceptually different (or different in type), but empirically equal in value. In this sense, gravitation is *universal* in Newtonian mechanics and obeys the weak equivalence principle (Carroll, 2004, pp. 48–49) – notably, without explanation. The geometrisation picture of general relativity, on the other hand, is often said to provide an explanation for the weak equivalence principle by eliminating m_g from the theory altogether, as we shall see below.

First, to better understand the geometrisation rationale and to prepare for the formulation of the *strong* equivalence principle let me rephrase the essence of the weak equivalence principle in a famous thought experiment (see Figure 3.1) by Einstein (2015, pp. 80–84), here summarised by Carroll:

Imagine … a physicist in a tightly sealed box, unable to observe the outside world, who is doing experiments involving the motion of test particles, for example to measure the local gravitational field. Of course she would obtain different answers if the box were sitting on the moon or on Jupiter than she would on Earth. But the answers would also be different if the box were accelerating at a constant velocity.

(Carroll, 2004, p. 49)

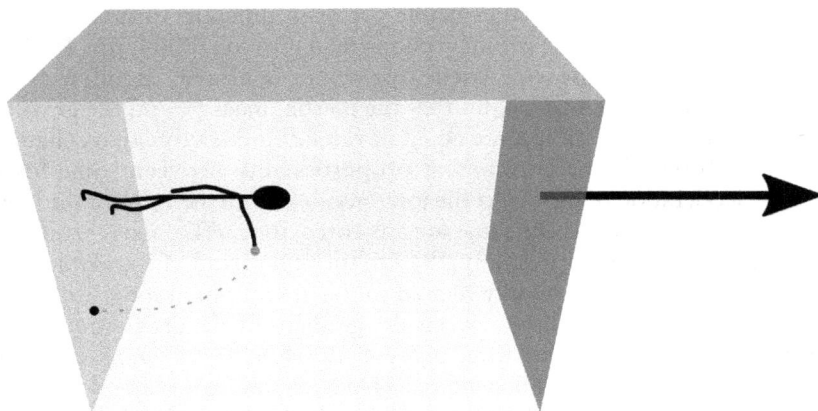

Figure 3.1 Einstein's elevator thought experiment. The behaviour of freely falling test bodies in a uniformly accelerated frame and in a gravitational field is the same.

Thus put, the weak equivalence principle states that it is impossible to distinguish whether the behaviour of freely falling test particles stems from a gravitational field or from their being situated in a uniformly accelerated frame. In contrast, it is possible to distinguish whether the behaviour of test particles stems from an electromagnetic field or from their being situated in a uniformly accelerated frame. This is because gravity is universal, whereas electromagnetism is not. With respect to the electromagnetic field, we would simply have to compare the behaviour of particles with different charge. Since all particles have the same "gravitational charge", this does not work (Carroll, 2004, p. 49).

It is important to note that the above holds for the case of a homogeneous gravitational field. If the gravitational field is inhomogeneous, then it only applies to sufficiently small frames; technically speaking, it only applies *locally*. We can then formulate a first version of the weak equivalence principle as follows:

> The motion of freely-falling particles are the same in a gravitational field and a uniformly accelerated frame, in small enough regions of spacetime.
>
> (Carroll, 2004, p. 49)

Since special relativity tells us that "mass" is just a manifestation of energy and momentum, a generalisation is in order (Carroll, 2004, pp. 49–50). Moving away from a restriction to "test particles" the strong equivalence principle generalises the above statement. It may tentatively be formulated as follows:

> In small enough regions of spacetime, the laws of physics reduce to those of special relativity; it is impossible to detect the existence of a gravitational field by means of local experiments.
>
> (Carroll, 2004, p. 50)

This means that locally we can always "transform away" a gravitational field and the laws reduce to the laws of special relativity. (Again, keep in mind that the current presentation and Carroll's definitions lack in precision (see Read et al., 2018 and Brown & Read, 2016, pp. 331–332). I shall correct some of the inaccuracies in Section 5.2.3. For the time being, I am more interested in giving a first and rough overview.) In this sense, gravity becomes a "pseudo force". There is no such thing as a gravitational potential in general relativity.

Now, this is not to say that gravity is fictitious. Quite the contrary, it means that gravity turns out to be *inescapable*: a "gravitationally neutral object" with respect to which we could measure the acceleration due to gravity *does not exist* (Carroll, 2004, p. 50). Hence, every object in the universe carrying energy and momentum is subject to gravity. In fact, every object is subject to gravity *in the same way*. Gravity does not distinguish between different types of objects. All objects, regardless of their properties including mass, are attracted *universally* (Carroll, 2004, p. 48); gravitational effects between test bodies only show in their relative acceleration.

It is precisely gravity's universality that seems to strongly suggest the geometrisation picture of gravity. For gravitation essentially *being* curvature of spacetime, *being* a feature of the metric field of a pseudo-Riemannian manifold, *being* a geometrical background structure nicely explains why the weak and the strong equivalence principle should hold. If gravitation is curvature of spacetime, then it seems obvious why we can always perform local transformations such that gravitation vanishes, and why the laws of physics locally look like the laws of special relativity. It then also seems obvious why this should affect every single object in the universe in the same way. The simple fact that gravitational effects are apparently independent of the objects' properties supports the claim that gravitation arises from spacetime itself and that the notion of gravitational mass needs to be eliminated. While gravitation *is* spacetime (or an aspect thereof), the other fundamental interactions are fields *in* spacetime.

As a result, the strong equivalence principle does not only play an important role for general relativity as a conceptual foundation of the theory, but also for the theory dualism in physics: a geometrisation picture of gravity seems fairly disconnected from how we understand the other fundamental interactions.

3.4   The Field View[23]

However, this perspective on general relativity, according to which g is, first and foremost, a geometrical entity – the spacetime metric – that only additionally "gives rise" to gravitation, is not exclusive. Lehmkuhl (2008) argues that general relativity does not commit us to the reduction of gravitation to spacetime geometry. Certainly, general relativity associates gravitation with spacetime geometry, but the type of association is not fixed by the theory (Lehmkuhl, 2008, p. 84). Besides geometrisation one may as well put forward a *field interpretation* – or, as a third option, what Lehmkuhl dubs an *egalitarian interpretation*.

The field interpretation essentially turns around the geometrisation picture. According to Lehmkuhl (2008) it was first introduced by Reichenbach:

> It has occasionally been said that this conception deprives gravitation of its physical character and that gravitation therefore becomes geometry. ... We are ... reversing the actual relationship if we speak of a reduction of mechanics to geometry: *it is not the theory of gravitation that becomes geometry, but it is geometry that becomes an expression of the gravitational field*. The theory of relativity did not convert a part of physics into geometry. On the contrary, even more physics is involved in geometry.
>
> (Reichenbach, 1958, p. 256)

So the field view takes spacetime geometry to be an aspect of, or to be reduced to what is considered a, first and foremost, *gravitational* field – the g field. Whereas this perspective seems to be more of a rhetorical point here, the field view becomes manifest in the case of the spin-2 approach to general relativity (see Chapter 4). Put in a slogan, g is "just another field" (like, for example, the electromagnetic field). It is only due to its specific properties that g also defines the geometry of spacetime. I come back to this in a moment.

The "distinction between viewing either geometry or gravity as more fundamental, and the idea of one being a manifestation or consequence of the other" (Lehmkuhl, 2008, p. 85) is supplemented by Lehmkuhl's third option – egalitarianism – which argues for a conceptual identification of gravity and spacetime in general relativity. In its strongest version, this is understood to have ontological import: gravity and spacetime are "two names for one and the same 'thing'" (Lehmkuhl, 2008, p. 84). Hence, one might suspect that the question of whether gravity is geometrised, or rather spacetime "gravitised", is largely a debate about semantics – as the egalitarian would arguably conclude.

Now, one reason for preferring the field view to geometrisation is that the field view emphasises the similarities between general relativity and the other field theories,[24] which may be considered relevant for unification (cf. Chapter 6). Some caution is advisable, though. Formal similarities between theories do not entail substantial and physical reasons for unification. Consider the case of Newton's law of gravitation and Coulomb's law of electricity: the fact that both laws exhibit the exact same mathematical form does by no means imply that the phenomena of gravitation and electricity are linked in any substantial sense.

Besides unification, there is a more important issue regarding the relation of general relativity to other theories to be considered here, namely the question of how the g field – or, in fact, any metric field – obtains its property to act as a metric that is surveyed by material rods and clocks in the first place (notably, regardless of the geometrisation–gravitisation debate). Famously, Harvey Brown and Oliver Pooley have raised this issue in their work on the dynamical approach, which can be seen to adopt a perspective of the field view type. But before I get to this, it is interesting to note that Steven Weinberg has pointed out an important caveat in advocating a "nongeometrical" (Weinberg, 1972, p. viii) understanding of general relativity – as he does himself: it comes with an additional explanatory burden regarding the equivalence principle (which is also relevant with respect to the dynamical approach). So let me take a step back and briefly review how Weinberg conceives of the whole situation.

Weinberg is explicit in expressing his opposition to the geometrical understanding of general relativity:

> In learning general relativity, and then in teaching it …, I became dissatisfied with what seemed to be the usual approach to the subject. I found that in most textbooks geometric ideas were given a starring role, so that a student who asked why the gravitational field is represented by a metric tensor, or why freely falling particles move on geodesics, or why the field equations are generally covariant would come away with an impression that this had something to do with the fact that space-time is a Riemannian manifold.
>
> (Weinberg, 1972, p. vii)

Weinberg regards the prevalence of the geometrisation view as a historical contingency that is rooted in the mathematical formalism which Einstein used originally. But "this historical fact does not mean that the essence of general relativity necessarily consists in the application of Riemannian geometry to physical space and time" (Weinberg, 1972, p. 3) – as posited by the geometrisation view, which thereby confuses, Weinberg argues, "representation" and "represented". Weinberg puts forward an attitude that may be summarised as: "Don't look at the formalism, look at the physics!" – arguably, reminiscent of Stachel's (1993, p. 17) charge against what he dubs "mathematical fetishism", or Reichenbach's comment on the geometrical interpretation's being "merely the visual cloak [*Gewand*]" (Reichenbach, 1928, pp. 353–354); see also Giovanelli (2021, p. 243).[25] Against geometrisation, Weinberg proposes to conceive of Riemannian geometry as a mere "mathematical tool" (Weinberg, 1972, p. viii) to account for "the peculiar empirical properties of gravitation, properties summarised by Einstein's Principle of Equivalence of Gravitation and Inertia" (Weinberg, 1972, p. 3).

To make it clear that what he regards as a mere mathematical tool is not to be confused with the physical content of the theory, Weinberg then turns the standard textbook presentation of general relativity around:

> In place of Riemannian geometry, I have based the discussion of general relativity on a principle derived from experiment: the Principle of the Equivalence of Gravitation and Inertia. ... so that Riemannian geometry appears only as a mathematical tool for the exploitation of the Principle of Equivalence, and not as a fundamental basis for the theory of gravitation.
>
> (Weinberg, 1972, p. viii)

However – this is crucial – after withdrawing from the natural explanation of the equivalence principle in terms of spacetime geometry, Weinberg needs to present an alternative explanation for it – as he is aware of:

> This approach naturally leads us to ask why gravitation should obey the Principle of Equivalence.
>
> (Weinberg, 1972, p. viii)

I come back to this question in the next section and, in particular, in Section 5.2.3. Notably, Weinberg does not expect to find an answer within the general framework of classical physics or within general relativity. Instead, Weinberg argues that one has to consider "the constraints imposed by the quantum theory of gravitation" (Weinberg, 1972, p. viii). In the course of Chapters 4 and 5, we shall see what this means, how this reasoning can be spelled out, and to what extent it is correct. But, first, let me move on and present how Brown and Pooley challenge the standard view in their dynamical approach.

3.5   The Dynamical Approach[26]

As we have seen, in both special and general relativity, the metric field (plus the manifold) is standardly taken to be fundamental and to represent spacetime. This is the fundamentalist view. In the case of general relativity, it gives rise to (or is supported by) the geometrisation picture of gravity: spacetime – or, more precisely, the curved metric field g – tells matter how to move. Underlying the claim that gravity is a manifestation of spacetime curvature is the more general claim that in any type of spacetime it is the spacetime geometry (represented by the respective metric field) which explains the motion of free material bodies.

There is another sense in which spacetime can be viewed to be explanatory: spacetime, or rather its metrical properties, may explain certain aspects about the *dynamical properties of the matter fields that build these material bodies* (e.g., the electron field, the quark fields, the photon field, and the gluon fields).[27] It is, first and foremost, the explanation of this second type which gives rise to what is usually dubbed the *geometrical* understanding of spacetime in both general and special relativity. It is what we have encountered previously as the claim that the dynamical symmetry properties of the matter fields coincide with (or are adapted to) – in the directed sense of "are explained by" – the symmetry properties of the metric (recall the fundamentalist's commitment (C) in Section 3.2). This geometrical view is contested by the dynamical approach, which was originally proposed by Brown (2005) and Brown and Pooley (2001, 2006).

3.5.1   *Spacetime: Explanans or Explanandum?*

But let me first reconsider the geometrical explanation of the first type regarding free inertial motion of test bodies. The empirical basis for the claim that spacetime is explanatory of free test body motion is that free test bodies in Newtonian, Minkowskian, or general-relativistic spacetime behave in a specific way that is related to properties of the spacetime: they move along the geodesics of the respective spacetime. But, begins Brown his investigation, "[w]hat is geometry doing here – codifying the behaviour of free bodies in elegant mathematical language or actually explaining it" (Brown, 2005, pp. 23–24)?

The fundamentalist opts for the latter. *Prima facie*, taking spacetime as explanatory seems to amount to taking the geodesics as "ruts or grooves in space-time which somehow guide the free particles along their way" (Brown,

2005, p. 24). This is arguably reminiscent of Hermann Weyl's (1970, p. VI) *Führungsfeld* (guiding field). Similarly, Nerlich (1976) argues as follows[28]:

> without the affine structure there is nothing to determine how the [free] particle trajectory should lie. It has no antennae to tell it where other objects are, even if there were other objects … It *is because space-time has a certain shape that world lines lie as they do*.
>
> (Nerlich, 1976, p. 264; as cited in Brown, 2005, p. 24)

Brown concedes merely – "[o]f course, Nerlich is half right" (Brown, 2005, p. 24) – that "there is a prima facie mystery as to why objects with no antennae should move in an orchestrated fashion. That is precisely the pre-established harmony, or miracle" (Brown, 2005, p. 24). Brown then pushes back:

> it is a spurious notion of explanation that is being offered here. If free particles have no antennae, then they have no space-time feelers either. How are we to understand the coupling between the particles and the postulated geometrical space-time structure?
>
> (Brown, 2005, p. 24)[29]

Brown concludes that spacetime geometry is merely a *codification* of free body motion – essentially, because he deems the fundamentalist explanation not only obscure, but also redundant:

> At the heart of the whole business is the question whether the space-time explanation of inertia is not an exercise in redundancy. … It is non-trivial of course that inertia can be given a geometrical description, and this is associated with the fact that the behaviour of force-free bodies does not depend on their constitution: it is universal. But what is at issue is the arrow of explanation. The notion of explanation that Nerlich offers is like introducing two cogs into a machine which only engage with each other. It is simply more natural and economical – better philosophy, in short – to consider absolute space-time structure as a codification of certain key aspects of the behaviour of particles (and/or fields).
>
> (Brown, 2005, pp. 24–25)

This also underlines the close connection between explanatory and ontological aspects.[30]

The key issue, whether spacetime geometry is "codifying" (Brown, 2005, p. 23) certain properties of matter or "actually explaining" them (Brown, 2005, p. 24), reappears when considering the geometrical explanation of the second type. It is here where the dispute then condenses to the recent geometrical–dynamical debate, which is on investigating the relation between spacetime geometry and matter field *dynamics* (rather than the relation between spacetime geometry and inertial motion of test bodies). Arguably, the latter is explanatory of the former: the (quantum) matter fields build the (classical) test bodies. Hence, argues Brown, "the operational meaning of the metric [with respect to inertial motion of test bodies; my remark] is ultimately made possible by appeal to quantum theory" (Brown, 2005, p. 9) and the dynamical symmetry properties of the quantum fields.[31]

So the advocate of the dynamical approach seeks to argue for what Butterfield calls "Brown's moral": that chronogeometricity (i.e., the property of a field to act as the metric that is surveyed by rods and clocks built from matter fields) is a property which metrical structures have only "by dint of detailed physical arguments" (Butterfield, 2007, p. 320), or as Brown himself puts it, chronogeometricity "is not an intrinsic feature" (Brown, 2005, p. 151) of metrical structures.[32] So the advocate of the dynamical approach seeks to demonstrate that a given metric field is *not in virtue of itself* surveyed by matter fields (or rather by the rods and clocks built from them), but *in virtue of further specifications*, namely in virtue of the (empirical) fact that the ontologically independent dynamical symmetry properties of the matter fields happen to coincide with the symmetry properties of the metric field. According to the dynamical approach, chronogeometricity is a *non-fundamental extrinsic property* of the metric field: chronogeometricity ontologically depends on the symmetry properties of the metric and the dynamical symmetry properties of the matter fields. If God had explicitly created these ontologically independent symmetry properties so that they match, then she would have implicitly created the metric's chronogeometricity. On the contrary, the advocate of the geometrical approach seems to conceive of chronogeometricity as something that a given metric field has *in virtue of itself*; a given metric field somehow explains the dynamical symmetry properties of the matter fields without further specifications (e.g., regarding their dynamics) and thereby secures that the symmetries match so that this metric field is surveyed by the matter fields (or rather by the rods and clocks built from them). If God had explicitly created the metric field and all its properties, then she would have implicitly created the dynamical symmetry properties of the matter fields[33]; arguably, it is this aspect that strikes the advocate of the dynamical approach as odd.

Consider special relativity, for example. One can ask whether it is the Minkowskian nature of spacetime that explains "why the forces holding a rod together are Lorentz-invariant or the other way around" (Balashov & Janssen, 2003, p. 340). For the proponent of the geometrical approach, spacetime geometry "is the explanans here and the invariance of the forces the explanandum" (Balashov & Janssen, 2003, p. 340). According to the geometrical view, the fundamental Minkowski metric somehow *determines* that the matter fields have Lorentzian dynamics, such that the Minkowski metric can be surveyed by material rods and clocks; the dynamical symmetry properties of the matter fields are non-fundamental (ontologically dependent on the symmetry properties of the Minkowski metric). But, according to Brown, "it is wholly unclear how this geometrical explanation is supposed to work" (Brown, 2005, p. 134). Instead, the proponent of the dynamical approach to special relativity proposes to turn around the explanatory relation between the symmetries of spacetime and the symmetries of the matter field dynamics (cf. Balashov & Janssen, 2003): the ontologically independent dynamical symmetry properties of the matter fields explain the symmetry properties of the Minkowski metric, which are hence deemed non-fundamental.[34]

This ultimately results in the following two central claims of the dynamical approach (see Brown & Read, 2021; Read, 2019): (a) the Minkowski metric $\eta_{\mu\nu}$ in special relativity is completely ontologically reducible to (or, supervenes on; see Norton, 2008, p. 822) the symmetry properties of the dynamics of matter fields; its chronogeometricity then follows automatically; and (b) the chronogeometricity of the metric field $g_{\mu\nu}$ in general relativity is a non-fundamental, extrinsic property, which g has in virtue of the strong equivalence principle, or, equivalently, in virtue of the symmetry properties of the matter field dynamics. I review these claims in more detail below.

Hence, the dynamical approach to relativity theory can be considered a philosophical variant of the field view. Against the geometrical understanding of general relativity, the field view takes the metric field in general relativity as – from the outset – a physical field like the others; or as the proponent of the dynamical approach puts it:

> the dynamics of the metric field tell us that it is 'just another field': "Nothing in the form of the equations *per se* indicates that g_{ab} is the metric of space-time, rather than a $(0, 2)$ symmetric tensor which is assumed to be non-singular" (Brown, 2005, p. 160).
> (Read et al., 2018, p. 18)

It is only in virtue of the equivalence principle that the g field then receives its status as the metric. Note, however, that g's being "a $(0, 2)$ symmetric tensor which is assumed to be non-singular" (Brown, 2005, p. 160) still makes it a candidate metric field, unlike a generic tensor, vector, or scalar field, for example.

Note that the dynamical approach to general relativity can be considered less successful than the dynamical approach to special relativity, because it only renders the g field's chronogeometricity non-fundamental;

the other properties of g (including other spatiotemporal properties, like being *a* metrical structure) remain fundamental. Recall that a field's having metrical properties is not the same as a field's having chronogeometricity: having metrical properties means to be *a* metric field, while having also chronogeometricity means to be *the* metric field which is actually surveyed by material rods and clocks. Hence, we may say that g is only partly reducible to matter field dynamics (unlike η), or that the fundamental g field has the non-fundamental extrinsic property of chronogeometricity. Thus, with regard to showing that spatiotemporal structure is derivative, the dynamical approach to general relativity is less successful than the dynamical approach to special relativity.

One need not take the reduction claim as the central claim of the programme, though. There is a sense in which the issue of whether a given metric has chronogeometricity is what the dynamical approach is after primarily, whereas the reduction claim in the case of special relativity is secondary; in this understanding, the failure to completely ontologically reduce the metric of general relativity is irrelevant for the programme to stand.

However, in trying to bring forward an explanation for g's chronogeometricity in general relativity, the proponent of the dynamical approach has to countenance two unexplained "miracles", as James Read, Harvey Brown, and Dennis Lehmkuhl argue in Read et al. (2018). In special relativity, on the contrary, only one miracle arises, precisely due to the complete ontological reduction of metric structure to matter field dynamics. I shall therefore take the reduction claim as central for a full-fledged dynamical approach, and, hence, view the dynamical approach to general relativity as less satisfactory – i.e., as "stuck" – compared to the dynamical approach to special relativity. A solution to this shortcoming of the dynamical approach to general relativity is presented in Section 4.2.2.

Focusing on the reduction aspect is in line with Norton's (2008) understanding of the dynamical approach as a constructivist programme that is directed against the standard fundamentalist conception of spacetime. It also draws attention to the fact that the geometrical–dynamical debate is closely related to the traditional substantivalism–relationalism debate. While the geometrical view is the standard interpretation of general relativity in terms of *explanation*, spacetime fundamentalism is the standard interpretation of general relativity in terms of its *ontological commitments*. The spacetime constructivist – *contra* the fundamentalist – holds that spacetime (geometry) is fully reducible to properties of matter fields. Indeed, at least in the dynamical approach to special relativity, the spacetime metric is derivative on the dynamics of the matter fields – which, in traditional terminology, is a form of relationalism, as Norton (2008) and Pooley (2013) agree.

In the following, I first properly introduce the dynamical approach to special and general relativity. I then address what Read et al. (2018) refer to as the appearing of "two miracles" in the dynamical approach to general relativity.

3.5.2   *The Dynamical Approach to Special Relativity*

In the case of special relativity with its fixed metric (see Pooley, 2017), the two opposing positions in the geometrical–dynamical debate are usually characterised as follows:

> We ... distinguish two positions on the nature of the Minkowski metric field:
>
> (A)   The Minkowski metric field is an ontologically distinct and primitive entity; its presence can explain certain facts about the dynamical laws governing matter fields (namely, the fact that these laws are Poincaré invariant).
> (B)   The Minkowski metric field is not an ontologically distinct and primitive entity; rather, it is a codification of certain facts about the dynamical laws governing matter fields (namely, the fact that these laws are Poincaré invariant).
>
> To endorse (B) is to endorse the dynamical perspective on relativity; the orthodox line is (A).
>
> (Read et al., 2018, p. 18)

I propose to reformulate the positions (A) and (B) as the geometrical (GA-SR) and the dynamical (DA-SR) approach to special relativity, respectively:

GA-SR   (1)   The Minkowski metric field η is an ontologically independent entity;
         (2)   it explains why (or determines that) the dynamical symmetry properties of the matter fields are Lorentz-invariant (or more precisely: Poincaré-invariant).

DA-SR   (1)   The Minkowski metric field η is *not* an ontologically independent entity,
         (2)   but is completely ontologically reducible to the Lorentzian symmetry properties of the matter field dynamics.

Friedman (1983), Maudlin (2012), and Norton (2008) hold some version of GA-SR; Brown and Pooley (2001, 2006) and Brown (2005) originally proposed DA-SR.

Typically, what is called the debate between the geometrical and the dynamical approach is a debate about *explanation*. The proposal of the dynamical approach is then often phrased as "reversing the arrow of explanation" with respect to the geometrical approach, because DA-SR takes the (symmetry properties of the) matter fields to do the explanatory work: "[t]he appropriate structure is Minkowski geometry *precisely because* the laws of physics of the non-gravitational interactions are Lorentz covariant" (Brown, 2005, p. 133); see also Read et al. (2018, p. 18) and Read (2020, p. 183). Phenomena like length contraction of a physical rod, for example, are not explained by properties of spacetime, but solely by the dynamical symmetry properties of the matter fields that build the rod: "a moving rod contracts, and a moving clock dilates, *because of how it is made up and not because of the nature of its spatio-temporal environment*" (Brown, 2005, p. 8). In this sense, "[o]ne of the guiding intuitions behind the dynamical approach concerns explanatory priority" (Pooley, 2013, p. 569). In short, the dynamical approach to special relativity maintains that the spacetime structure of a world that is described by special relativity is Minkowskian *because of the dynamical symmetry properties of the matter fields that build physical bodies*.

As a result, the dynamical approach also expresses a novel *ontological* position. DA-SR does not just reverse the arrow of explanation, but opposes special-relativistic spacetime fundamentalism, which upholds that η is ontologically independent, i.e., fundamental. The explanatory priority of the dynamical symmetry properties of the matter fields justifies the ontological claim that the dynamical symmetry properties of the matter fields are *more fundamental* than the spacetime structure, which is a derivative entity. In this sense, the dynamical approach to special relativity is a novel form of relationalism (Pooley, 2013, Section 6). The Minkowski metric's being derivative means that it is ontologically dependent, i.e., non-fundamental.

More concretely, the Minkowski metric is completely ontologically reducible to properties of matter fields, namely their dynamical symmetry properties. This is because the Minkowski metric is a rather simple structure: a fixed field without its own dynamics that is characterised *only* by its Lorentzian symmetry properties; this was precisely the reason for why Minkowski relationalism is easily achievable (see Section 3.2). Due to this complete ontological reduction, one can even argue that the Minkowski metric is not only a non-fundamental entity, but also, paraphrasing Brown (2005, p. 9) and Read (2020, p. 183), no more than the *codification* of these dynamical symmetry properties of the matter fields, i.e., a "glorious non-entity" (Brown & Pooley, 2006); it is not part of the ontology at all. Note, however, that Martens (2019) could again argue that since the Lorentzian symmetry properties of the matter fields are real (i.e., in the ontology), the Minkowski metric is as well; "it is just not fundamental" (Martens, 2019, p. 11). Accordingly, Martens would arguably view the Minkowski metric as a non-fundamental, redundant, but nevertheless existing *entity* (*pace* Brown and Pooley); similar to how the classical rods and clocks themselves are non-fundamental, redundant, but nevertheless existing entities, which are, thus, part of the ontology (although not part of the fundamental ontology). For my specific project of answering the question whether spacetime is *fundamental* (here, in special relativity), it is already sufficient to defend the weaker claim that the Minkowski metric is non-fundamental and to refrain from additionally arguing whether it should be banished from the ontology altogether.

To summarise, the dynamical approach to special relativity states that the spacetime structure of a world that is described by special relativity is non-fundamental, because it is *completely determined by the dynamical symmetry properties of the matter fields*. Hence, DA-SR qualifies as a type of relationalism:

> The dynamical approach seeks to offer a reductive account of the Minkowski spacetime interval in terms of the dynamical symmetries of the laws governing matter. It therefore qualifies as a type of relationalism, although this is not something that Brown himself emphasises.
>
> (Pooley, 2013, p. 569)

More precisely, Pooley introduces the dynamical approach to special relativity – alongside Huggett's (2006) regularity relationalism for Newtonian mechanics – as a version of "have-it-all relationalism" (Pooley, 2013, p. 564), where "some of the spacetime structure implicit in the dynamics is judged to have only an effective status, ultimately grounded in a less structured relationalist ontology" (Pooley, 2013, p. 564).

3.5.3   The Dynamical Approach to General Relativity

In the case of general relativity, the debate is different, due to the g field's being dynamical. In particular, with respect to ontology the proponent of the dynamical approach is prepared to agree with the geometrical (or fundamentalist) view that the g field is *not* derivative on properties of matter fields, but fundamental; or, as Read, Brown, and Lehmkuhl put it:

> The first key difference in the case of GR [general relativity] is that the advocate of … [the dynamical approach] concedes that the metric field in this context is an autonomous agent, ontologically distinct [read: ontologically independent; my remark] from the matter fields of the theory. Hence, for the extensions to GR of *both* … [the geometrical approach] and … [the dynamical approach], the metric field is *not* reducible to properties of the matter fields.
>
> (Read et al., 2018, p. 18)

Similarly, Read (2020) summarises that in general relativity

> advocates of both the dynamical and geometrical approaches agree that the metric field g_{ab} is an ontologically autonomous entity [read: ontologically independent entity; my remark], obeying its own dynamical equations, and not straightforwardly reducible to (symmetries of dynamical equations governing) matter fields, as per the dynamical approach to SR [special relativity].
>
> (Read, 2020, pp. 183–184)

It is on the issue of chronogeometricity where the two camps are considered to still disagree. The question how the ontologically independent g field is equipped with its property to act as the metric that is surveyed by material rods and clocks is the key issue for the dynamical approach to general relativity. Roughly, the geometrical approach takes g to have its chronogeometricity *intrinsically*, i.e., in virtue of itself, while the dynamical approach takes g to have its chronogeometricity *extrinsically*, namely in virtue of the specific form of matter field dynamics (determined by the strong equivalence principle), as a non-fundamental property that is derivative on the symmetry properties of g and the dynamical symmetry properties of the matter fields. In summary, the two camps maintain the following (see, for example, Read et al., 2018):

GA-GR   (1)   The g field is ontologically independent.
                 (2)   It has its chronogeometricity intrinsically, i.e., in virtue of itself.

DA-GR   (1)   The g field is ontologically independent.
                 (2)   It has its chronogeometricity extrinsically, namely in virtue of the strong equivalence principle.

So in contrast to DA-SR – where chronogeometricity follows from the metric field's being determined by the symmetry properties of the matter field dynamics – DA-GR needs to invoke an additional postulate, namely the strong equivalence principle. This is because "nothing in the form of the equations *per se* indicates that $g_{\mu\nu}$ is the metric of space-time" (Brown, 2005, p. 160), rather (see also Read, 2020)

> it is because of … local Lorentz covariance that rods and clocks, built out of the matter fields which display that symmetry, behave as if they were reading aspects of the metric field and in so doing confer on this field a geometric meaning.
>
> (Brown, 2005, p. 176)

For the dynamical approach, it is via the strong equivalence principle that g acquires its chronogeometricity[35]:

> chronogeometrical significance of the $g_{\mu\nu}$ field is not an intrinsic feature of gravitational dynamics, but earns its spurs by way of the strong equivalence principle.
>
> (Brown, 2005, p. 151)

On the other side, Read et al. (2018, p. 19) take proponents of GA-GR to typically argue that "the metric field has a primitive connection to spacetime geometry": it is "the existence of the Lorentzian metric field" which "*explains* the form of the local dynamical laws in the theory" (Read et al., 2018, p. 19); chronogeometricity is *intrinsic* to g. The explanatory work in GA-GR is done by the fundamental g field alone. Since g is argued to have its chronogeometricity intrinsically in GA-GR,[36] the strong equivalence principle is not considered to do any additional explanatory work. Note that this does make for a minor ontological dissent between the two camps: chronogeometricity has a different ontological status and the g field has a different set of properties.

To summarise, the proponent of the geometrical approach takes chronogeometricity as a property that g has in virtue of itself, i.e., as an intrinsic property of g. The proponent of the dynamical approach, on the contrary, takes chronogeometricity as a property that g does not have in virtue of itself, but rather in virtue of matter field properties that are determined by the strong equivalence principle, i.e., as an extrinsic property of g – arguing that "nothing in the form of the equations *per se* indicates that $g_{\mu\nu}$ is the metric of space-time" (Brown, 2005, p. 160). Indeed, the strong equivalence principle does not follow from the Einstein field equations, but needs to be introduced independently by hand (e.g., Brown, 2005; Lehmkuhl, 2021).

This is the underlying key argument of the proponent of the dynamical approach. For a given candidate metric field to have chronogeometricity, i.e., for it to be *the* actual metric that is surveyed by material rods and clocks, the dynamical symmetry properties of the matter fields, which build these rods and clocks, must match the symmetry properties of this candidate metric field. So, for g to be the metric field of general relativity that is surveyed by material rods and clocks, the dynamical symmetry properties of the matter fields, which build these rods and clocks, must match the symmetry properties of the g field. The symmetry properties of g are fixed by the Einstein field equations to be locally Lorentz-invariant (or, more precisely, locally Poincaré-invariant). However, the corresponding dynamical symmetry properties of the matter fields are *not* fixed by the Einstein field equations themselves, but by the strong equivalence principle,[37] which is an additional assumption. So, asks the proponent of the dynamical approach, if the dynamical symmetry properties of the matter fields are not determined by the Einstein field equations, then why should g have its chronogeometricity intrinsically? The proponent of the geometrical approach effectively assumes that the content of the strong equivalence principle is inscribed to (or explained by) g, which would make it inscribed to (or explained by) the Einstein field equations, *which it is not.*

This favours the dynamical approach over the geometrical approach to general relativity. It is by raising certain problem cases for the geometrical approach that Read et al. (2018) and Read (2020) make this explanatory deficit of the geometrical approach precise (e.g., by considering Galilean matter field dynamics). Read (2020) refines this rationale and significantly clarifies the debate by distinguishing two versions of the geometrical approach – the *unqualified geometrical approach* (UGA), which views chronogeometricity as *intrinsic* to g, and the *qualified geometrical approach* (QGA), which accepts that g has its chronogeometricity only in virtue of certain additional assumptions, i.e., *extrinsically*. Both QGA and UGA are geometrical approaches in the sense that they affirm the ontologically independent status of candidate metric structure in a given theory. As Read (2020) remarks, it is primarily UGA that is attacked by Brown (2005), Brown and Pooley (2001, 2006), and Read et al. (2018). The position that I presented above as "the" geometrical approach, is to be identified with UGA.

In contrast to UGA, QGA claims that, given a particular set of dynamical equations for matter fields, the dynamical behaviour of these matter fields is explained by the candidate metric field *because in the particular set of dynamical equations under consideration these matter fields couple to this metric field.* On QGA, the metric has chronogeometricity *only because of the additional specification of the matter field dynamics*, i.e., *not* in virtue of itself, but in virtue of the matter field dynamics. Therefore, QGA accepts the chronogeometricity charge by the dynamical approach (Read, 2019, p. 104), i.e., QGA accepts that chronogeometricity is an *extrinsic* property of the metric field that depends on certain additional assumptions – which, in the case of general relativity, are summarised by the strong equivalence principle.

Read (2020) then argues that only QGA is tenable due to certain problem cases, which essentially draw on the fact that the Einstein field equations do not prohibit locally non-Lorentzian matter field dynamics. It just is perfectly consistent with the Einstein field equations that the matter field dynamics do not manifest the symmetries of g. For instance, without further input assumptions "there is nothing to rule out dynamical equations for non-gravitational fields in GR which are locally *Galilean* invariant" (Read, 2019, p. 105).

Finally, Read (2020) concludes that the tenable (qualified) version of the geometrical approach and the dynamical approach are explanatorily on a par (both must accept the same brute facts). In fact, he takes it that any distinction between the dynamical and the tenable geometrical approach to general relativity collapses:

> while an untenable form of the geometrical approach may purport to account for both MR1 and MR2 [the two miracles (see next section)], any acceptable form of the geometrical approach must also accept these two miracles of GR. In this sense, the existence of these two miracles is independent of the dynamical/geometrical debate.
>
> (Read, 2020, p. 185)

For special relativity the situation is different: the dynamical approach and QGA are distinguished, on the one hand, by their ontological claims, and, on the other hand, by the fact that the dynamical approach – unlike QGA – rigorously explains chronogeometricity (with accepting fewer brute facts) via the ontological reduction of the metric field. This stresses once again that the dynamical approach to general relativity is less satisfactory than that to special relativity.

3.5.4   Two Miracles

The issue that trying to explain chronogeometricity potentially involves accepting certain unexplained brute facts can be made more precise by distinguishing between: (1) the fixing of a universally shared dynamical symmetry property for all matter fields (expressing the core statement of the strong equivalence principle; see Section 5.2.3), and (2) the coinciding of this shared dynamical symmetry property with the symmetry property of the metric field (thereby conveying chronogeometric significance to the metric field). Accordingly, Read et al. (2018, p. 20) identify the following two potential "miracles" (i.e., unexplained brute facts) in the foundations of relativity theory[38]:

MR1   All matter field dynamics are locally Poincaré-invariant.

MR2   The local symmetry properties of matter fields coincide with the local symmetry properties of the metric field.

For special relativity, the dynamical approach accepts only MR1 as unexplained – indeed, MR2 receives an explanation through the complete ontological reduction of the Minkowski metric. For general relativity, the dynamical approach accepts both MR1 and MR2 as unexplained brute facts. In contrast, the geometrical approach (or, more precisely, its unqualified variant with which the works prior to Read (2020) primarily take issue) "attempts to rationalise" (Read et al., 2018, p. 20) both miracles by means of the respective ontologically independent metric field, which is argued to appropriately "constrain" the dynamics of the matter fields, i.e., to explain the miracles (see also Read, 2020). However, such an explanation must seem question-begging from the perspective of the dynamical approach, or so Read, Brown, and Lehmkuhl argue:

> One thing that the advocate of (A) [the (unqualified) geometrical approach] may say here is the following: Minimally coupled dynamical laws in GR [general relativity] feature the metric field g_{ab}; as we have seen, the presence of (or rather, the coupling to) *this metric* constrains the local form in the neighbourhood of any $p \in M$ of the dynamical laws of those fields to which it couples. Consequently, the symmetries of the local dynamical laws *must* coincide with the symmetries of the metric field. This argument misses the point, however, for the very issue in question is why the dynamical laws governing matter fields take such a form – rather than another, with *different* local symmetry properties. In other words: *why this particular coupling?* This is the essence of MR2, which remains untouched by such arguments.
>
> (Read et al., 2018, p. 20)

The concrete problem cases which Read et al. (2018) present demonstrate that the explanatory work cannot be done by the metric field alone – contrary to what is claimed by the (unqualified) geometrical approach. Absent further argument, the position is indeed not tenable and the charges on what Read later dubs UGA do find their mark: UGA is to be dismissed on the basis of the problem cases; without further input assumptions, the metric field g does not suffice to obtain MR1 and MR2. It just is perfectly consistent with special relativity and general relativity that the matter field dynamics do not manifest the symmetries of the respective metric field; after all, "matter enters through the energy-momentum tensor, which does not determine the equations of motion for matter, or their symmetries" (Huggett & Wüthrich, book manuscript, Chapter 9, p. 28). As already mentioned above, without further input assumptions "there is nothing to rule out dynamical equations for non-gravitational fields in GR [general relativity] which are locally *Galilean* invariant" (Read, 2019, p. 105). Notably, this is because the metric has been introduced as "self-standing" and "autonomous", i.e., as ontologically independent:

> As a matter of logic alone, if one postulates spacetime structure as a self-standing, autonomous element in one's theory, it need have no

constraining role on the form of the laws governing the rest of the theory's models. So how is its influence supposed to work? Unless this question is answered, spacetime cannot be taken to explain the Lorentz covariance of the dynamical laws.

(Brown & Pooley, 2006, p. 84)

This argument is accepted by any tenable variant of the geometrical approach (i.e., QGA). It is only with the help of further input assumptions, which restrict the dynamics of matter fields to a certain form, that the metric field can feature in explanations of matter field dynamics. Accordingly, QGA accepts both miracles as unexplained brute facts – not only in general relativity but in special relativity as well (Read, 2020; see also below).

According to Read et al. (2018) and Read (2020), all this suggests the following conclusion: the unexplained existence of MR1 and MR2 is *independent* of endorsing a particular view in the geometrical–dynamical debate. It is a fact about general relativity itself that MR1 and MR2 do not receive an explanation, not a fact about a particular interpretation. Regarding MR2, this is due to the following. Recall that the

> key difference in the case of GR [general relativity] is that ... the metric field in this context is an autonomous agent, ontologically distinct from the matter fields of the theory. Hence, ... the metric field is not reducible to properties of the matter fields.
>
> (Read et al., 2018, p. 18)

Due to the apparent irreducibility of g the second miracle appears:

> if one could argue that what had previously been regarded as the ontologically independent metric field of GR [general relativity] was in fact reducible to a codification of symmetries of the dynamical equations of matter fields (as ... in SR [special relativity]), then this would provide an explanation for MR2.
>
> (Read et al., 2018, p. 20)

This is why Read (2019) turns to string theory to seek the explanation of the two miracles that general relativity is not able to provide. I come back to these and related issues regarding the dynamical approach and the interpretation of general relativity in Chapter 4. In what follows, based on joint work with Niels Linnemann, I first discuss another problem that has been directed at the dynamical approach by Norton (2008), the *problem of pregeometry*.

3.6 The Problem of Pregeometry[39]

As we have seen, the dynamical approach draws attention to the fact that metric structure is, at least in part, ontologically dependent on matter field structure. According to the dynamical approach to special relativity, the Minkowski metric is derivative on (i.e., ontologically reduced to) symmetry properties of matter field dynamics; according to the dynamical approach to general relativity, at least the g field's chronogeometricity ontologically depends on matter field properties. So there is a sense in which the project of Harvey Brown and Oliver Pooley's dynamical approach to relativity theory can be labelled "constructivist". Recall that spacetime constructivism is the doctrine that spatiotemporal structure is derivative on matter field dynamics, or, as Norton puts it, that "spacetime theories are essentially matter theories" (Norton, 2008, p. 821) which are "devoid of spatiotemporal presumptions" (Norton, 2008, p. 825).[40] Accordingly, spacetime constructivism is opposed to spacetime fundamentalism.

3.6.1 *Preliminaries*

Norton (2008) takes this up and argues that the dynamical approach fails to accomplish the constructivist programme: it does not render *all* spatiotemporal structure derivative on matter field dynamics. This failure of the dynamical approach is apparent for general relativity, but Norton maintains that the dynamical approach to *special* relativity fails as well. Setting up matter fields, Norton argues, requires some kind of topological background structure, a manifold. Supposedly, this topological background structure is spatiotemporal and part of the fundamental ontology. Call it *pregeometry*. As a result, on the dynamical approach *there are* "components of spacetime geometry that do not supervene on matter and also cannot be explained by matter theories; that is, they would not, in the language of Brown's text, be a 'result of' properties of matter" (Norton, 2008, p. 823) – contrary to what a constructivist understanding of the dynamical approach seeks to establish. Instead, or so Norton claims, "the construction project must tacitly assume an already existing spacetime endowed with topological properties" (Norton, 2008, p. 824).

Norton exposes that Brown and Pooley as well as other proponents of the dynamical approach are (almost) solely concerned with the ontological status and explanatory role of the metric. They do not disclose what ontological status and explanatory role they ascribe to the *manifold*, although, according to orthodoxy, spacetime is represented by the *pair* of manifold structure and metric structure. Thus, the proponent of the dynamical approach is urged to clarify their position.

The proponent of a *truly constructivist* dynamical approach is also challenged to show that the pregeometry is either (a) non-fundamental (e.g., derivative on the field dynamics), (b) non-spatiotemporal (the pregeometry does not bring in spatiotemporality), or (c) non-ontic (the pregeometry is not part of the overall ontology at all, i.e., not even of the non-fundamental ontology, but a purely representational artefact). Either of these options would ensure that Norton's concern is moot. Call this challenge on the way to full spacetime constructivism the *problem of pregeometry*.

Note that the general perception is that Brown and Pooley themselves do *not* advocate spacetime constructivism (see Norton, 2008; Pooley, 2013), but are only concerned with linking chronogeometricity to matter field dynamics. Accordingly, they, and other proponents of such a *non-constructivist* dynamical approach, might simply bite the bullet and accept that their fundamental ontology includes the pregeometry (*contra* full constructivism). It is interesting to note, though, that Brown occasionally does seem to have in mind something more ambitious, as remarks in the context of the hole argument demonstrate (similarly in Brown, 1997, p. 68):

> if the bare, differential space-time manifold is a real entity, then different solutions of Einstein's field equations that are related by diffeomorphisms correspond to different physical states of affairs. The theory is incapable of predicting which of the different possible worlds is realized, but all of them are, as we have seen, empirically indistinguishable. *The simplest (and to my mind the best) conclusion, and one which tallies with our usual intuitions concerning the gauge freedom in electrodynamics, is that the space-time manifold is a non-entity.* In this case the different, diffeomorphically related worlds are not only observationally indistinguishable, they are one and the same thing.
>
> (Brown, 2005, p. 156; emphasis added)

In the following, we investigate whether a constructivist take on special and general relativity can succeed with respect to the ontological status of the manifold. We first present Norton's challenge for the constructivist (see Section 3.6.2), and map out possible replies at a general level (see Section 3.6.3). We then assess and criticise the only explicitly proposed solution strategy, namely that by Menon (Section 3.6.4). Dissatisfied with this response to Norton, we provide two independent lines of defence of spacetime constructivism of our own: the first strategy is suggested by an account of the dynamical approach due to Stevens (Section 3.6.5), the second rejects Norton's view of the manifold as an indispensable object (Section 3.6.6).

3.6.2 *What Is the Problem of Pregeometry?*

Let us start by introducing in more detail the problem of pregeometry for the proponent of the dynamical approach who wants to endorse spacetime constructivism. Take the constructivist to claim – contrary to the standard fundamentalist conception of spacetime – that *all* spatiotemporal properties are *fully* reducible to matter field dynamics; in particular, that even topological properties are fully reducible to matter field dynamics. The problems of pregeometry for the proponent of the dynamical approach who wants to endorse constructivism then take the following forms for special and general relativity (PoP-SR and PoP-GR, respectively):

PoP-SR   The dynamical approach ontologically reduces the Minkowski metric to matter field dynamics. The dynamical approach fails to reduce core spatiotemporal properties that are not specifically associated to metric structure but which are required to set up the matter fields in the first place.[41] These core spatiotemporal properties are expressed in the assumption of a fundamental *joint* manifold structure, called pregeometry.

PoP-GR   The dynamical approach renders the chronogeometricity of the g field – a core spatiotemporal property – as ontologically dependent on the matter field dynamics.[42] The dynamical approach fails to reduce, amongst others, core spatiotemporal properties which are required to set up the fields in the first place. These core spatiotemporal properties are expressed in the assumption of a fundamental *joint* manifold structure, called pregeometry.

As a consequence, the dynamical approach cannot be considered full-fledged constructivist – call it *half-way constructivist*. Take the case of special relativity: in response to Norton, Brown and Pooley – not pursuing the project of reducing all spatiotemporal structure but merely metric structure to the dynamical properties of the matter fields – admit that they do presume a pregeometric differentiable manifold structure (Norton, 2008, p. 829; Pooley, 2013, pp. 572–573).[43] To count as full-fledged constructivist, the dynamical approach would have to render *all* (putatively) spatiotemporal structure – not only metric structure – derivative on the dynamics. As the metric structure in special relativity already is derivative on the dynamics, this essentially amounts to either rendering manifold structure derivative on the dynamics as well, or establishing why manifold structure should not count as spatiotemporal in the first place. Neither option can succeed, Norton argues: "[t]he construction project must tacitly

assume an already existing spacetime endowed with topological properties, so that it can introduce spatiotemporal coincidences, and a unique
set of standard coordinates (x, y, z, t)" (Norton, 2008, p. 824). Here is
a way to understand Norton's worry: suppose, for instance, that all we
have are two matter fields ϕ_1 and ϕ_2 each defined on manifolds M_1 and M_2,
respectively. We still expect our theory to answer a question of the following form: "The value of field ϕ_1 at a point P_1 in the manifold M_1 is $\phi_1(P_1)$,
what is the value of field ϕ_2 at this same point?" To be able to answer such
a question, we seem to need to assume that M_1 and M_2 are identical, such
that P_1 and P_2 are identical (see also Figure 3.2).[44] In a sense, the proponent
of the dynamical approach has to deal with a "third miracle"[45]:

> Nothing in the matter theories individually assures us that the stand-
> ard coordinate systems (x_3, y_3, z_3, t_3) of matter theory 3, say, and the
> standard coordinate systems (x_7, y_7, z_7, t_7) of matter theory 7 are the
> same.
>
> (Norton, 2008, p. 829)

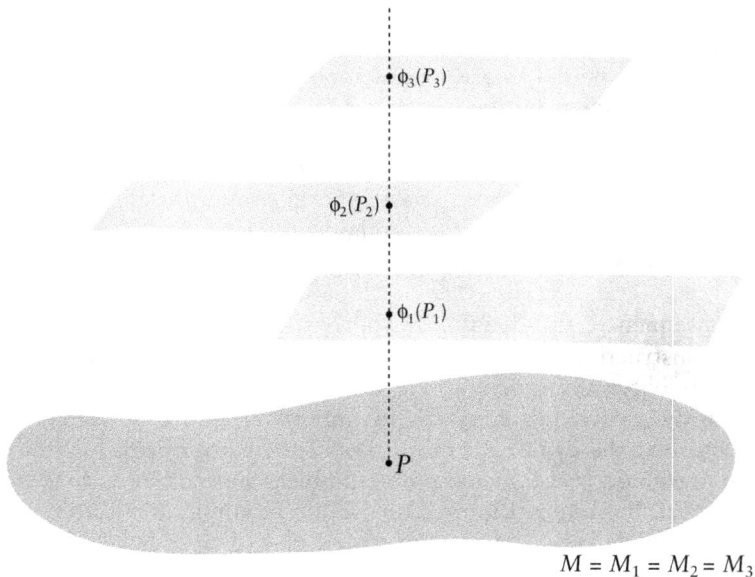

$$M = M_1 = M_2 = M_3$$

Figure 3.2 Visualisation of Norton's indispensability argument for the manifold
(dubbed here the "pincushion rationale"). Given some matter fields ϕ_i
defined on individual manifolds M_1, M_2, M_3, respectively, only identifying all individual manifolds with the single manifold M (the "pincushion") allows to compare field values of different fields at a point P.

Put differently, Norton argues that measurements obtained through rods and clocks built from different matter fields ϕ_i will only then return the same spatiotemporal distance Δs between two events A and B, if the parameters (x_i, y_i, z_i, t_i) used for each matter field ϕ_i can be taken to *coincide*, i.e., to describe the same points on a joint underlying manifold structure. This essentially is "the supposition of the standard coordinate systems (x, y, z, t) of the realist conception" (Norton, 2008, p. 829) – see the first claim of Norton's "realist" conception of spacetime in Section 3.2.

Now, before we get to the details of this debate, suppose Norton is right. Suppose the dynamical approach does have to assume a pregeometry that is (a) fundamental, (b) spatiotemporal, and (c) ontic. One might ask: what is wrong with this half-way constructivism? Such a version of constructivism may seem "unfinished" and unsatisfactory (e.g., if some form of relationalism was to be defended), but maybe that is just how the world is. The constructivist could simply double down on the original dynamical approach, accept Norton's charge that some spatiotemporal structure is presumed, but argue that this "half-way" approach is novel and interesting nonetheless. Half-way constructivism paints a different ontological picture than the standard fundamentalist conception, after all; it may, for example, be cashed out as some form of spacetime emergentism. We shall briefly come back to this option of biting the bullet later. It is important to point out, however, that we do take Norton's claim to be controversial. And we aim to demonstrate that the dynamical approach is *not* committed to accept the manifold as part of the ontology.

3.6.3 *Replies to the Problem of Pregeometry*

Recall that Norton claims that the dynamical approach needs to presume the manifold as a spatiotemporal part of the fundamental ontology. Accordingly, the constructivist is challenged to show that the manifold is, in fact, either (a) non-fundamental (e.g., derivative on the matter field dynamics), (b) non-spatiotemporal (it does not bring in spatiotemporality), or (c) non-ontic (it is not part of the overall ontology, but a purely representational artefact or "scaffolding structure"). Either of these options would adequately respond to the challenge.

We will consider the following strategies for solving the problem of pregeometry: (1) showing that the manifold is derivative on the matter field dynamics; (2) arguing that the manifold should not count as spatiotemporal; and (3) arguing that the manifold is dispensable auxiliary structure.

Route 1 can be read as the attempt to establish that the manifold should never have been counted as "pre"geometry in the first place, because it is just as well derivative on the matter field dynamics. The manifold is ontologically reducible. This strategy is discussed in Section 3.6.4. Route 2 seeks to demonstrate that the manifold does not bring in spatiotemporality. This can be read as the attempt to establish that the manifold should never have been counted as pre"geometry". The strategy requires a clear conception of non-spatiotemporality as opposed to spatiotemporality. We discuss this option in Section 3.6.5. Finally, route 3 argues that the manifold is dispensable or non-ontic auxiliary structure. This option comes in two variants: (a) the manifold is dispensable (we really only use it because it is convenient) and, thus, non-fundamental in the determination sense, or (b) the manifold is – albeit *in*dispensable in the formal set-up – *dispensable* from the actual physical set-up, i.e., non-ontic. In both variants, the manifold does not appear in the fundamental ontology. Notably, this route is partly concerned with realism (see Section 3.6.6).

What has been dubbed half-way constructivism above does not follow either of these routes, but simply denies that there is a problem to begin with. We do acknowledge that half-way constructivism is a tenable position that has the advantage of being independent of additional philosophical assumptions regarding spatiotemporality or ontological status. When adopting half-way constructivism, one does not see Norton's pregeometry challenge as threatening the dynamical approach as such, but as merely demanding the dynamical approach to clarify its stance on the manifold.

3.6.4   *Debunking Menon's Algebraic Approach*

For his solution to the problem of pregeometry, Menon (2019) uses the algebraic reformulation of general relativity going back to Geroch (1972). Fields including the g field are standardly represented as structures on a manifold. If this manifold is then interpreted as a fundamental, spatiotemporal part of the ontology, this amounts to presupposing a pregeometric structure and Norton is right. Now, the algebraic approach seems to challenge this interpretation. Here, the basic idea is to use the rich algebraic structure of the fields (that is how they can be combined, i.e., multiplied, added, etc.) to, in the end, do away with the manifold structure at the fundamental level and make it derivative on the field dynamics.

For reasons of simplicity, we will – similarly to Menon – only consider scalar fields in the following.[46] Then, the set of infinitely differentiable real-valued fields $C^\infty(M)$, usually understood as a set of fields on a manifold, can be reconceptualised as the abstract algebra underlying it. The manifold version is then just a specific representation of this algebra; notably, using less than this (abstract) algebraic structure does not suffice (see Menon, 2019, p. 1277).

Menon interprets this as follows: he takes the (abstract equivalent of a) field to be fundamental on the algebraic view (as an element of an algebra, and thus as an abstract entity). In contrast, the manifold points are reconstructed from these abstract entities and thereby rendered *derivative*. If this holds, one could say that the manifold is actually ontologically re-ducible to a non-spatiotemporal algebraic structure in terms of which the dynamics are expressed fundamentally. Hence, the problem of pregeom-etry would be solved.

Unfortunately, that is a bit too fast. Using the well-known theoretical equivalence between the manifold and the algebraic representation (see Rynasiewicz, 1992; Rosenstock et al., 2015) we now argue that either the "how much pregeometry needed" question (1) only sets up a pseudo-problem (it is no problem in the algebraic representation because it is no problem in *any* representation, i.e., it is not a problem at all), or (2) poses a genuine problem that cannot be transformed away through a mere change of representation. In other words, all Menon aims to do in the algebraic representation can already be done – if at all – in the standard manifold representation, or so we now argue.

First, recall that Norton claims that "[t]he construction project must tacitly assume an already existing spacetime endowed with topological properties, so that it can introduce spatiotemporal coincidences, and a unique set of standard coordinates (x, y, z, t)" (Norton, 2008, p. 824). In our reading, Menon essentially understands this claim as a two-step chal-lenge: (1) show that it is even possible to construct, for example, special relativity *without* assuming a joint base structure, and (2) show that a joint base structure is not required to introduce spatiotemporal coincidences. Menon tries to address these issues by showing that, regarding (1), there is, in fact, the option of starting from *several* base structures, and, regarding (2), one can still obtain spatiotemporal coincidences.

Here is Menon's argument in more detail. He begins by distinguishing between kinematically possible models (KPMs) of two kinds at the level of manifold structure:

> KPMs of relativistic theories are tuples of the form $\langle M, g_{ab}, \phi_i \rangle$, where M is a smooth manifold, g_{ab} is a Lorentzian metric tensor, and ϕ_i is a placeholder for matter fields. …
>
> Models of this kind implicitly … have an extra bit of structure – a link between the manifold and the fields. Specifically, for any point in M, distinct fields, $\phi_1, …, \phi_n$, can be 'evaluated' at that point, and their values can then be taken … to represent properties of the same space-time point. Call a KPM in which this link does exist a *KPM of the first kind*.
>
> …
>
> Given the arbitrariness of the coordinate function, there is no a priori need for the coordinatization of two composite functions to coincide. In other words, there is no a priori reason to require that

the x in $\phi(x)$ is the same as the x in $\psi(x)$ – indeed, our use of the same variable to denote both is indicative of a failure to allow for this mismatch. Call models in which preimages of x in $\phi(x)$ and $\psi(x)$ can represent different points, *KPMs of the second kind.*

(Menon, 2019, pp. 1274–1275)

So, KPMs of the first kind presuppose a joint base structure ("fields are defined on one and the same structure") whereas KPMs of the second kind, at least *prima facie*, do not ("fields can be defined on completely independent structures").

KPMs of the first kind are naturally linked to a manifold representation (see quote). Now, Menon's intuition is that KPMs of the second kind are *not* linked to a manifold representation. On this picture, showing KPMs of the first kind to be derivative on KPMs of the second kind would then show that the manifold is derivative, too. As Menon points out "[t]he central aim of this paper is to demonstrate that KPMs of the second kind can be constructed for SR [special relativity]" (Menon, 2019, p. 1275), and that "Norton's criticism is based on assuming the KPMs at play can only be of the first kind" (Menon, 2019, p. 1275). But we have just seen that even Menon himself shows that KPMs of the second kind *can* be constructed in the manifold representation: we just need to accept that we then have multiple distinct manifolds, i.e., one for each field, in our KPMs. So, even if Menon's story about how to obtain the KPMs of the first kind from the KPMs of the second kind in the algebraic formulation is correct, an exact counterpart can be told in the manifold representation.

Essentially, the point is that the equivalence of the manifold and the algebraic representation makes Menon's move a mere change of representation, which cannot be ontologically relevant. Or to put it differently, the equivalence of the manifold and the algebraic viewpoint implies that neither view is ontologically privileged. Neither view is ontologically more relevant than the other. Both manifold and algebraic representation seem to go along with their own (natural or formalistic) interpretation but the formal equivalence prevents any prioritisation. In fact, it may even suggest a joint interpretation. Thus, some additional reason is needed as to why one interpretation should be preferable over the other – and, arguably, even to demonstrate that there should be two distinct as opposed to one joint interpretation of the two equivalent representational structures. Without further argumentation, it is not clear how mere re-representation solves the ontological problem at stake (or any ontological issue to begin with).

In addition, Norton's very criticism is *based* on considering the option of there being KPMs of the second kind in the manifold representation in the first place: arguing that we can accept several distinct manifolds for special relativity does not tell against Norton's key argument that we are forced to accept a single manifold structure as the result of a common origin inference.

But there is a subtlety regarding the re-representation argument against Menon. What we have rejected seems oddly familiar: usually, the dynamical approach proponent is prepared to *deny* that a formal equivalence between two statements prevents prioritisation or suggests a joint interpretation.[47] After all, the basic idea behind the dynamical approach to, say, special relativity is to think of the metric as derivative, *although*, purely mathematically speaking, stating that equations of motion display Lorentz invariance is equivalent to stating that there exists a fundamental Minkowski metric. In this case, the dynamical approach proponent does prioritise one of two formally equivalent views. It would be rather inconvenient, if our argument against Menon were to refute the dynamical approach altogether. Fortunately, both cases differ. Here, one can adhere to a basic posit of the dynamical approach, namely "explanations should be dynamical not geometrical". Menon's argument cannot build on a similar argumentation – at least not from the dynamical approach perspective.

At the same time, simply saying that the algebraic view is preferable over the manifold view because the former is non-spatiotemporal, is a *petitio principii*; the dynamical approach proponent must show why her view leads to full constructivism, not argue that it is compatible with it.[48]

We do acknowledge, however, that Menon (indirectly) gives an account of under which circumstances one can assume coincidence of coordinates, that is, in other words, assume a single ordering structure for all fields:

> Consider, now, a coordinate system on which we have good reason to believe that, say, a Gaussian wave packet of the $\phi(x)$ field bounced off a Gaussian wave packet of the $\psi(x)$ field in the neighborhood of some point. There will be a class of coordinatizations [i.e., functions from M to $\mathbb{R}^n$; our addition] that assign field values in such a way that the dynamics of each field determines that a collision took place in the vicinity of some point and another class of coordinatizations on which the kinks in the trajectories (or, more generally, some fact about the dynamical interaction) of each particle do not take place at the same coordinate value. For various reasons, the former class of coordinate systems might be preferable. For free fields, there simply is no operational sense of spatiotemporal point coincidence, although such points can still be defined. In such a case, though, these would amount to arbitrary stipulations.
>
> (Menon, 2019, p. 1279)

One can only adhere to this explanation of point coincidences if one has already accepted a scenario of a multitude of manifolds, one for each field. It does not solve the problem of pregeometry, though. For this we still have to either dig deeper to find out how these various manifolds are derived from the dynamics (and, thus, even if spatiotemporal, are only derivatively so), or come up with some argument as to why the manifolds are all non-spatiotemporal (although fundamental). Either suffices to solve the problem of pregeometry. Given that the algebraic version of the first strategy just failed, let us consider the second.

At the outset, we observe the following: first, it seems natural to ascribe the same spatiotemporal status to all posited manifolds. Either all manifolds are equally spatiotemporal, or none of them is. Second, positing several spatiotemporal manifolds seems disadvantageous to positing a single spatiotemporal manifold.

The upshot then is the following. We may accept that our ability to assume multiple *non-spatiotemporal* manifolds is a solution to the problem of pregeometry *because* a commitment to multiple *spatiotemporal* manifolds would be implausible (whereas it is supposedly plausible to have multiple *non-spatiotemporal* manifolds). If there are multiple manifolds, they must be non-spatiotemporal. However, Norton can argue that manifold structure is always spatiotemporal and then insist that the commitment to a single (spatiotemporal) manifold is more plausible: if manifold structure is always spatiotemporal, it is better to assume a single spatiotemporal manifold than a plethora of those (which, again, is deemed implausible). In other words, the crucial problem is whether manifold structure *as such* is spatiotemporal or not (regardless of how many manifolds are considered); here, one will be able to draw on the proposal below that manifold structure lacks essential features of the spatiotemporal – in particular, a difference between space and time. In addition, with respect to Occam's razor type arguments, to thoroughly address Norton's insisting on a single manifold, the constructivist arguably has to show that the multiple non-spatiotemporal manifolds are non-ontic (i.e., merely representational). For, while Occam's razor does not exclude having several non-spatiotemporal and non-ontic ordering structures, it might exclude having several non-spatiotemporal but ontologically relevant ordering structures (and does certainly exclude having several spatiotemporal and ontologically relevant ordering structures – especially, when they exhibit point coincidences that allow for a common origin inference à la Janssen (2002a; 2002b) as Norton argues). At first sight, the option to posit several manifolds squares well with the constructivist claim that a manifold is mere representational (i.e., non-ontic), non-spatiotemporal ordering structure. On the other hand, the fundamentalist claim that manifold structure is fundamental and spatiotemporal will receive further support, if it is demonstrated that this claim is most plausible for a *single* manifold; this is exactly what Norton achieves with his common origin inference, aimed against the constructivist. If manifold structure was mere ordering structure, he argues, it would seem odd that the project is most successful when a *common* ordering structure is assumed. The success or plausibility of full constructivism should not depend on a specific choice for the ordering structure. Before we come back to this below, we shall now turn to two further solutions to the problem of pregeometry.

3.6.5   *The Problem of Pregeometry in Light of Stevens's Regularity Relationalism*

In this section, we argue that the conceptualisation of the dynamical approach by Stevens (2018) inspires a reply to the problem of pregeometry that is mostly in line with route 2, which seeks to demonstrate that the manifold does not bring in spatiotemporality. It must be clearly stressed that at no point Stevens himself makes reference to the pregeometry problem; in fact, he himself explicitly sees what we call pregeometry as "spatiotemporal sub-metrical structure" (Stevens, 2018, p. 357) and so apparently has not – at least at time of writing – taken up a non-spatiotemporal reading of the manifold which we take to be heavily suggested by his own account, or so we shall argue.

As presented above, the general task for the constructivist is to pull spacetime out of the dynamics governing the fields without already presuming spatiotemporal structure. For this, we can learn from Stevens and his analogy between the role of the Leibniz group in Newtonian mechanics and the role of the manifold structure in the dynamical approach. The analogy draws on the fact that in both contexts the fundamental setting has less structure than the actual dynamical setting of the theory: in the case of a Leibnizian relationalism on Newtonian mechanics, there are full Leibnizian symmetries at the fundamental ontological level, while at the level of the actual dynamics only Galilean symmetries reside. Thus, the less symmetric dynamical structure is supposed to supervene on the more symmetric fundamental structure (more symmetry means less structure).[49] In the case of the dynamical approach to special relativity, there are matter fields on a manifold *without metric* at the fundamental ontological level, while at the actual dynamical level all matter fields move in a Minkowski spacetime, i.e., a manifold *with metric*.

More concretely, Leibnizian spacetime – unlike Galilean spacetime – is invariant under linear accelerations, i.e., has no affine structure. Accordingly, Newtonian theory cannot be straightforwardly interpreted in terms of Leibnizian relationalism (see also Section 2.2.1) – the dynamical and spatiotemporal symmetries do not match. But, as Pooley (2013) points out, the Leibnizian relationalist does have the following three options to resolve this issue. In order to accommodate for more structured (less symmetric) dynamics, the Leibnizian relationalist could either (1) extend her ideology, that is the set of relations between the fundamental entities of the theory, or (2) move to a different theory committed to different fundamental entities, or (3) seek "a way to reconcile restricted dynamical symmetries with more permissive spacetime symmetries", i.e., show how to "have one's cake and eat it" (Pooley, 2013, p. 564). The general idea of option (3) is that at least part of the spatiotemporal structure has "only an effective status, ultimately grounded in a less structured relationalist ontology" (Pooley, 2013, p. 564).

A well-known example for option (3) is Huggett's (2006) *regularity relationalism*, where the dynamical laws of Newtonian Mechanics described in Galilean spacetime supervene on a Humean mosaic in Leibnizian spacetime; not only the actual dynamics but *also* that it is described in Galilean spacetime arises as a liberalised Humean best-systems analysis. Since Leibnizian spacetime is fundamental, Huggett's approach is not a fully constructivist project either. But due to the analogy between option (3) and the general programme of constructivism, the lessons of Huggett's account can be put to use in the pregeometry debate when trying to have spacetime supervene on non-spatiotemporal structure (i.e., route 2).

Now, with Huggett's proposal for Leibnizian relationalism at hand, Stevens (2018) cashes out the aforementioned analogy between Leibnizian relationalism and the dynamical approach by arguing that the pregeometric structure is simply that structure with respect to which the patterns of a Humean mosaic are formulated. Concretely, a successful constructivist project would take a material world with the pregeometric structure as the supervenience basis and then best-systematise the dynamical laws in terms of a (more) spatiotemporal structure that supervenes on it (Stevens, 2018).

At first sight, this supervenience basis might be conceived of as a fundamental *spatiotemporal* supervenience basis. As a matter of fact, Stevens himself explicitly speaks of "spatiotemporal sub-metrical structure" (Stevens, 2018, p. 357).

Our point now is simple but forceful: rendering pregeometry as the complete supervenience basis *for everything else* is tantamount to making the pregeometric structure independent of any field that might be added (or, put technically, defined) on top of it. So on this account the pregeometric structure is a fundamental, i.e., ontologically independent entity.[50] But once this is realised, the question immediately becomes as to how one could understand a mere manifold structure as spatiotemporal. First, it is essential to space and time (or spacetime) to play the role of an ordering structure. If the different concepts of time from physics to psychology and phenomenology have anything in common, it is the idea that time, amongst other things, is an ordering parameter. The same holds for space. Second, it is essential to space and time that the two are in a relevant sense distinct from one another.[51] Given this view on the spatiotemporal, the manifold cannot count as spatiotemporal: *it is an ordering structure but lacks a distinction between one ordering parameter as opposed to the others.*

3.6.6 The Problem of Pregeometry as a Problem of Indispensability

Let us retrace our steps. Menon approaches the problem of pregeometry via route 1: he tries to establish manifold structure as non-fundamental (i.e., dependent) by making it derivative on the algebraically formulated dynamics. We have argued that Menon's proposal is ill-directed. We have then pointed out that Stevens's conceptualisation of the dynamical approach can be read to inspire a solution to the problem of pregeometry similar to route 2: showing that manifold structure may be fundamental, but non-spatiotemporal. We have argued that this solution depends on a concrete notion of the spatiotemporal. A third option remains to be discussed: arguing that the manifold is dispensable or non-ontic auxiliary structure (route 3).

Recall that Menon's account entails that fields, at first defined on distinct manifolds, can be taken to refer to one and the same manifold through considering their dynamical evolution; the different fields are not free, but interact. This result addresses Norton's *coincidence concern*, namely that "[t]he construction project must tacitly assume an already existing spacetime endowed with topological properties, so that it can introduce spatiotemporal coincidences, and a unique set of standard coordinates (x, y, z, t)" (Norton, 2008, p. 824).

Note, however, that Norton (2008, p. 828) explicitly disputes that field interactions can explain spatiotemporal point coincidences. We find his counter argument which considers a toy model of a distance-based coupling term hardly convincing, but shall pursue a different line of argumentation. Nevertheless, only interactions make the coincidences operationally accessible (Menon, 2019, p. 1279). If all we had were free fields, spatiotemporal coincidences "would amount to arbitrary stipulations" (Menon, 2019, p. 1279). This disarms Norton's final conclusion on the matter, namely that "interactions are of no help if we consider two fields that do not interact" (Norton, 2008, p. 828). Also note that the gravitational interaction is universal.

Figuratively speaking, one can dub Norton's position the "pincushion rationale" (see Figure 3.2): evaluating the field values of different fields at a point involves fixing these fields to one and the same background structure reminiscent of a needle's fixing different layers of fabric to a pincushion. As a consequence, the rationale claims that one is committed to (what serves as) the "pincushion" as part of the fundamental ontology – this is Norton's indispensability argument for the manifold.

However, committing to the "pincushion" yields too much: it is the "*needles*" (i.e., the field dynamics) that fix the coincidences, not the "pincushion" which additionally fixes an absolute localisation by associating a spacetime point to each "needle". When all we need is coincidences, as Norton argues himself, why should we be forced to commit to the "pincushion"? Thus put, another way to block Norton's coincidence concern over and above that proposed by Menon is – or, is at least motivated by – what we call the "needles only rationale". Speaking less figuratively, what ultimately is at stake is whether or not the coincidence concern translates to an indispensability argument for the *manifold of spacetime points* specifically, and whether the manifold's alleged ontological status as well as its being spatiotemporal follow from it.

Notably, Brown (1997) similarly remarks that

> [i]t is tempting to assign such autonomy ['reality' in Brown (2005, p. 156); our remark] to the continuum of space-time points precisely because a dynamical field is standardly defined as a map from this continuum into a suitable set of tensors of a certain kind. The temptation is closely related to the intuition which led Lorentz to maintain the notion ... of an imponderable ... ether, whose ultimate role was to provide "a peg to hang all these things [electromagnetic fields] on". Lorentz's ether failed in the minds of most physicists to survive the developments of 1905, and in particular Einstein's persuasive demonstration of the "redundancy" of the electromagnetic medium, ponderable or otherwise. Yet arguably the core of Lorentz's intuition was still widely, if unwittingly, adhered to: the "peg" ... now became the space-time manifold itself.
>
> (Brown, 1997, p. 68)

Brown then goes on to stress, once again, that the debates in the context of diffeomorphism invariance and the hole argument in general relativity call "into question the reality of the post-Lorentzian peg. In pre-quantum physics then, space-time points are perhaps best viewed not as entities in their own right, but as correlations or links between the individual degrees of freedom of distinct physical fields" (Brown, 1997, p. 68). Arguably, this is a variant of our "needles only rationale" which, here, is bolstered by the observation that we usually do not commit to gauge redundancies as physically real (see also Brown, 2005, p. 156).[52] We take it that this marks an important sense in which the coincidence concern does *not* translate to an indispensability argument for the manifold specifically.

Now, the "needles only rationale" seems to be easily countered by the insight that the "totality of needles", or the totality of what Brown calls "correlations" or "links" (Brown, 1997, p. 68), may just reflect the manifold structure in disguise, so that nothing really is gained by this argument. However, or so we insist, it reflects the manifold structure only in so far as that the manifold provides some representational structure for the *physical* events: manifestly, the "totality of needles" is the totality of (interaction) events, *not* the totality of spacetime points; it is a manifold of "needles" or correlations, not a manifold of spacetime points (the latter of which, again, would additionally fix an absolute structure that is *only afterwards* argued to bear a certain redundancy in virtue of diffeomorphism invariance and the hole argument).[53] Each "needle" can be seen to refer to a manifold point, but manifestly, the "needles" only mark correlations between different field values. For instance, a supernova event gives physical correlations between the gravitational, the electromagnetic, and other fields (like neutrino fields) that can be used to fix their relative displacement in a certain neighbourhood. We can associate such correlations of gravitational, electromagnetic and other physical events with a particular point of the manifold of spacetime points. The manifold of spacetime points is (just) a convenient way to organise this information.

In fact, note that for this to work we are not even committed to a *manifold* of "needles": not every event needs to be an interaction of different matter fields (a "needle" event) and not all manifold points need to correspond to events, we just need an adequate (possibly discrete) network of interaction events such that we can coordinate the different field structures.[54] As a result – *pace* Norton – the manifold is, indeed, *dispensable* for the explanations Norton is after. This points to the non-fundamentality of the manifold in the determination sense: no additional information is needed, all that needs to be determined is determined by the correlations. The manifold is dispensable from the fundamental ontology. Notably, independently of whether the manifold is dispensable from the most convenient formalism or shown to be derivative, i.e., dependent. Hence, the standard concepts of fundamentality seem to come apart here: the manifold may not be needed to determine, but could still be independent.

To bolster the dispensability stance further, one may seek arguments to establish the *physical* irrelevance of the manifold structure, i.e., that the manifold is non-ontic.[55] In particular, one can try to support our argumentation for the explanatory dispensability of the manifold of spacetime points (and, hence, its dispensability from the *fundamental* ontology) by refining the criteria for what counts as ontic or real as such (similar to Brown). Then, the mere fact that some structure features in our theories is insufficient for its featuring in the ontology. For instance, the fact that we view certain matter fields as real is not simply a result of their being part of our theories, but a result of their having *physical significance* (in a sense yet to be spelled out). The fact that we usually do *not* view gauge redundancies or auxiliary "ghost" fields as real is precisely in virtue of such additional criteria. Arguably, this is also in line with physics practise.

So, the general task is to argue for some version of what is sometimes dubbed *selective realism*. A selective realist denies that a sensible realist position amounts to accepting the reality of all ingredients of our well-established physical theories – some ingredients may merely be representational surplus structure.

Based on such an argument, we can then independently deny the ontological relevance of the pregeometric structure: pregeometry does not have physical significance, as it does not meet certain criteria – call this the *physical significance argument*. The argument by Brown (1997) regarding diffeomorphism invariance is of this kind. For example, one may link the physical significance – and thus the existence – of a theoretical entity to the *practise of physics* (implying that the formalism alone does not tell which entities or operations are physically significant). Curiel (2018) defends such a view when arguing against the existence of spacetime points:

> an entity ..., purportedly represented by a theoretical structure, has physicality if one has a reason to take that structure seriously in a physical sense, namely, if one can show that it plays an ineliminable or at least fruitful and important role in the way that theory and experiment make contact with each other.
>
> (Curiel, 2018, p. 472)

Curiel (2018, p. 476) also provides a list of criteria (e.g., having energy-momentum), at least some of which must be satisfied for a theoretical entity to be physically significant. However, with regard to the specific project of refuting Norton's indispensability argument, Curiel's considerations of physics practise do not suffice. This is because Norton's indispensability argument does *not* simply rely on the formalism, but already includes such practical considerations. What is more, it is Curiel who grants physicality to theoretical entities which play "an ineliminable (albeit physically obscure) role in the mathematical structure required to formulate the theory" (Curiel, 2018, p. 476), which, in the light of Norton's argument, is too permissive and, in fact, in danger of telling against Curiel's own conclusion that spacetime points are not physical. All hinges on the formalism again.[56] So dropping this criterion would be best: after all, it is typically in the very case that something plays an "ineliminable, albeit physically obscure, role" in the formulation of a theory that one wonders whether it should be considered physical or not. Thus, we should not accept that this criterion settles the question.

Finally, consider the following additional option to fuel constructivism: promising successor theories might indicate that pregeometry *will* turn out as non-fundamental (or dispensable); call this the *successor theory argument*. In fact, successor theories to relativity theory (or generally more encompassing theories) in the context of quantum gravity often do render the manifold even more as a physically insignificant scaffolding structure. This is true, for example, for candidate theories like loop quantum gravity. We may infer from their input that even if the manifold was indispensable in relativity theory, this is not robust over theory change. Hence, the constructivist may conclude that the manifold cannot be ontologically relevant; the successor theories have a say in what is part of the ontology.

However, while it might seem plausible that a quantum theory of gravity does away with substantival manifold structure, it is by no means true for all candidate theories. For example, the effective quantum-field-theoretic spin-2 approach does employ manifold structure (see Section 5.2). Also, one might object that such arguments have the downside that they heavily rely on speculative physics and, so to speak, the principle of hope. More importantly, to have some impact on the interpretation of the old theory, a rather specific continuity between the new and the old theory needs to be established as well. One may argue that not any kind of correspondence actually does the job.

With regard to Norton's challenge, however, the successor theory argument is off-target anyway. To engage in a debate about whether one can be a constructivist on the basis of a *different*, putatively "final" theory sidesteps the actual question.

3.6.7 Concluding Remarks

To summarise, we agree with Norton that the problem of pregeometry for the dynamical approach has generally received too little attention, despite its centrality. We have therefore appraised the only existing solution (besides half-way constructivism) and proposed two new ones.

We started out by evaluating Menon's proposal. Unfortunately, his account is not convincing without further justification as to why one of two equivalent representations is ontologically more relevant. This issue could be remedied by considering successor theories, on which also Menon (2019) ultimately seems to rely (see Huggett et al., 2021).

We then worked out two independent proposals of our own. The first is inspired by Stevens' conceptualisation of the dynamical approach as regularity relationalism: a best-systems account like that of regularity relationalism implies that the supervenience basis is independent of what supervenes; on such a view of the dynamical approach, the manifold – playing the role of the supervenience basis – must then be seen as ontologically independent of its field content. Whether or not the manifold is spatiotemporal, however, is a different question. Given that spatiotemporality minimally involves a distinction between space and time, the manifold structure – the pregeometry – can only be read as non-spatiotemporal.

The second proposal builds on the insight that what Norton essentially puts forward is an indispensability argument for the manifold (notably, on the basis of physical, not purely formal considerations). We countered that the manifold's indispensability is, in fact, *not* straightforwardly obtained: Norton's central argument only establishes the indispensability of point coincidences, which Norton then takes to imply the indispensability of the manifold. Against this, we argued that point coincidences do, in fact, suffice, if modelled in terms of interactions à la Menon – which is feasible independently of his algebraic approach. In addition, one may employ independent arguments against the ontological relevance of the manifold.

Which reply is preferable depends on (rather reasonable) background assumptions. We favour the dispensability account. The dispensability account of the manifold fits to how practitioners of quantum gravity think dismissively of the manifold. Conversely, when seen from the vantage point of the practitioner, the regularity relationalist account can be accused of interpreting the manifold too literally. What is more, background assumptions are less crucial in the case of the dispensability account. Certain assumptions about what counts as real (or physically significant) may come in handy, but the dispensability account has its intrinsic merits either way. On the other hand, however, the background assumptions about what counts as spatiotemporal will make or break the regularity relationalism account.

Let us add that we do acknowledge that half-way constructivism is a tenable option. When adopting half-way constructivism, one does not see Norton's pregeometry challenge as threatening the dynamical approach as such, but as merely demanding the dynamical approach proponent to clarify their stance on the manifold. Note, however, that adopting half-way constructivism arguably prompts the task of re-evaluating more closely whether half-way constructivism tells against the lessons drawn from the hole argument: if the hole argument and diffeomorphism invariance tell against the reality of the manifold, as Brown alleges himself, then they might put pressure on the tenability of half-way constructivism as well. This cautions us that half-way constructivism need not be a "safe bet".

3.7   Fundamentality of Spacetime Revisited

In this chapter, I studied what special and general relativity can tell us about the fundamentality of spacetime or, rather, certain spatiotemporal aspects. I argued that, according to the standard view, both theories seem to suggest that, for example, metrical aspects are fundamental, i.e., ontologically independent. I then showed how Brown and Pooley's dynamical approach to relativity theory begins to challenge this conclusion. On a dynamical view, metrical properties are argued to be ontologically *dependent* on matter field dynamics, thereby calling into question spacetime fundamentalism. However, assessing the dynamical approach further then reveals a couple of caveats. First, only the dynamical approach to special relativity is able to fully ontologically reduce metric structure to matter field structure, thus seriously challenging spacetime fundamentalism by proposing a form of relationalism. On the dynamical approach to general relativity, however, only the specific property that the g field acts as the metric that is surveyed by material rods and clocks is argued to ontologically depend on certain properties of the matter fields. So, the dynamical approach to general relativity merely demonstrates that its *chronogeometricity* is to be conceived of as ontologically dependent: g's chronogeometricity is determined by the symmetry properties of g and the dynamical symmetry properties of the matter fields; by *explicitly* creating matching symmetry properties for g and the matter fields, God would have *implicitly* created chronogeometricity as well. Hence, chronogeometricity is a non-fundamental and extrinsic property of the g field (g has this property only *in virtue of the matter fields*). Similarly, a massive body's weight is derivative on both its mass and the external gravitational field. However, the dynamical approach does not ontologically reduce the g field *as such* to matter field dynamics – everything else about g (e.g., its symmetry properties, its gravitational properties, and its having energy and momentum) is ontologically *independent*. This is precisely why chronogeometricity is an extrinsic property: since the symmetry properties of g are *intrinsic* properties that are ontologically independent of the (intrinsic) symmetry properties of the matter fields, g and the matter fields can have different symmetry properties, such that g does not intrinsically have chronogeometricity. Thus, the dynamical view fails to complete its programme of reducing spatiotemporal structure to non-spatiotemporal structure – it is stuck.

Note that this is not true if one conceives of chronogeometricity as being the only spatiotemporal property. One might argue that chronogeometricity is what distinguishes g from the other fields and makes it "the spacetime metric". Hence, one might argue, if g's chronogeometricity is non-fundamental because it ontologically depends on matter field properties, then there is nothing at the fundamental level that distinguishes any field as spatiotemporal. Thus, spatiotemporality is non-fundamental. Moreover, one may argue that the second miracle then only *seems* to arise because one mistakenly still holds on to viewing g as the "metric" field, despite the fact that g's being the metric field has just been shown to be a non-fundamental property of g. I do not adopt this view that chronogeometricity is all there is about spatiotemporality, nor that the second miracle thus trivialises. This is because there are other crucial aspects of spatiotemporality that may or may not be fundamental independently of the status of chronogeometricity. For example, g, whether fundamentally chronogeometric or not, may still determine the inertial motion of material test bodies, which is a widely acknowledged aspect of spacetime (e.g., Knox, 2019) that, according to the standard understanding of the weak equivalence principle, holds irrespective of the properties of these test bodies (but see also Section 5.2.3). In addition, as mentioned before, having chronogeometricity is not the same as being a metric field, as bimetric theories make especially clear; in bimetric theories there are two candidate metric structures, but only one of them has chronogeometricity.

Second, the dynamical view's being stuck is highlighted by the fact that the dynamical approach to general relativity has to accept *two* unexplained miracles, whereas the dynamical approach to special relativity only has to accept one (due to the complete ontological reduction of the Minkowski metric). Notably, Read (2020) provides reasons for why the unexplained appearance of these miracles affects both approaches (contrary to how the proponent of the standard geometrical approach perceives her position) and hence should not be viewed as supportive or obstructive to either the dynamical or the geometrical approach. Still, the issue does not increase confidence in the relevance of the general programme of the dynamical approach – especially considering that Read ultimately concludes that tenable variants of a geometrical and a dynamical view are, in fact, indistinguishable.

Third, we have seen that the dynamical approach is confronted with another problem, namely the problem of pregeometry. Although Niels Linnemann and I have provided reasons why this need not bother the dynamical approach, in sum, the worry might arise that the dynamical approach does not solve anything, or add any substantial insight to the standard view of spacetime, but rather creates problems.

Let me stress that I do not (fully) share these concerns. In particular, I do take it that the dynamical approach, despite its problems, points out – and partially solves – important issues in the philosophy of spacetime. For example, regarding special relativity, the dynamical approach is ontologically and explanatorily more convincing than the geometrical approach. Moreover, regarding general relativity, the dynamical approach forces the proponent of the geometrical approach to refine their position. Nevertheless, I regard the contribution of the dynamical approach to the question whether spacetime is fundamental as unsatisfactory. In particular, unlike the dynamical approach to special relativity, Brown and Pooley's dynamical approach to general relativity (which, ultimately, is more relevant) does not establish the non-fundamentality of metric structure, but rather gets stuck quite early on. In the following chapter, I aim to refute the resentments against the dynamical approach by showing how the dynamical approach to general relativity can push back the concerns it has to confront and thus be resurrected.

Notes

1 The following summary mainly follows Carroll (2004). Other standard text-books are, for example, Weinberg (1972), Misner et al. (1973), Wald (1984), and Maggiore (2008).

2 The Lorentz indices μ and ν run from 0 to 3 in integer steps. Thus, $g_{\mu\nu}$ is represented by a 4×4 matrix with 16 components. Since $g_{\mu\nu}$ is symmetric, only ten components are independent.

3 It is widely accepted that the alternative interpretation which identifies space-time with manifold structure alone – dubbed "manifold substantivalism" – is excluded by the hole argument (see Norton, 2019).

4 Some bimetric theories are now confronted with unfavourable experimental data from the binary neutron star merger observed by the LIGO-Virgo collaboration (Abbott et al., 2017). Essentially, since in some bimetric theories, electromagnetic and gravitational waves do not travel on the null geodesics of the same metric, one would generally expect different velocities. Thus, the "simultaneous detection of GW [gravitational waves] and EM [electromagnetic] signals rules out a class of modified gravity theories, termed 'dark matter emulators', which dispense with the need for dark matter by making ordinary matter couple to a different metric from that of GW" (Boran et al., 2018).

5 Philosophically, Bartels (2013; 2015, pp. 90–114) utilises the contingency of the relation between metric and connection in the context of the metaphysics of laws of nature to argue that according to general relativity metrical properties are not "powers" – if they were, this would imply that metrical properties cause the gravitational effects on matter with metaphysical necessity. Note that Bartels concedes to the dispositional monist that metrical properties in general relativity are fundamental (cf. Chapter 4) and dispositional, but contests that there is a metaphysically necessary determination of gravitational effects (e.g., tidal forces) by the metric g – the causal relation between the metric g and its manifestation, the tidal forces, is a nomological relation: it is determined by the particular choice of an affine connection (see endnote 7), i.e., by the geodesic principle, which – leaving a few qualifications aside (see Malament, 2012) – is entrenched in the Einstein field equations.

6 In fact, the distribution of matter and energy represented by $T_{\mu\nu}$ only *constrains* the curvature of spacetime, but does not fully determine it. This is because spacetime curvature is not represented by $G_{\mu\nu}$, but $R^\rho_{\sigma\mu\nu}$. For example, all components of the energy-momentum tensor vanishing is consistent with both spacetime being flat and being rippled by gravitational waves.

7 If the covariant derivative of a tensor T vanishes along some path $x(s)$, we say that the tensor is parallel transported along this path. A path that parallel transports its own tangent vector is called a geodesic. Given the standard choice of the metric-compatible Levi-Civita connection, this is equivalent to saying that a geodesic is the shortest path between two spacetime points – note that the concept of parallel transport depends on the choice of the affine connection.

8 This is the *geodesic principle* of general relativity that can be thought of as a relativistic version of Newton's First Law of Motion (Malament, 2012, p. 245). While the geodesic principle and the geodesic equation are often presented as a direct consequence of the Einstein field equations, Malament (2012) draws attention to the fact that a few other assumptions are involved as well.

9 For a translation see Einstein (1923).

10 Brown (2005) emphasises that inertial frames are global in special relativity. In general relativity the relativity principle is restricted to arbitrary local coordinate transformations. In this sense general relativity is indeed a restriction rather than a generalisation of special relativity: "it cannot be emphasized too strongly that the latter [general relativity] is from a certain point of view not at all what its name seems to indicate; it is indeed from a certain point of view not a generalization but rather a restriction of the so-called Restricted Theory [special relativity]. It restricts its validity to the infinitesimal neighbourhood of any – and that means, of course, of every – world-point" (Schrödinger, 1950, p. 82; see also Brown, 1997, p. 68).

11 "Henceforth space by itself, and time by itself, are doomed to fade away into mere shadows, and only a kind of union of the two will preserve an independent reality" (Minkowski, 1923, p. 75). For the original publication see Minkowski (1909).

12 See also Fletcher (2020).

13 Viewing the metric as not representing spacetime may be justified by the fact that the metric field of general relativity gives rise to gravity and, in particular, gravitational waves – which arguably carry energy and momentum (see Earman & Norton, 1987; Read, 2018; but cf. Hoefer (1996) and Dürr (2019)). This may be viewed as putting the metric field into the matter category (e.g., Martens, 2019; see also Section 2.2.4).

14 Norton stresses that the "realist conception" could, of course, also be formulated in terms of the manifold M and the Minkowski metric η (referring to (1) and (2), respectively).

15 I shall prefer to speak of "matter fields" and "matter field dynamics" rather than "matter theories" to emphasise the ontological viewpoint.

16 Naturally, the spatiotemporal relations are the intervals $\Delta s = \sqrt{c^2 \Delta t^2 - \Delta x^2 - \Delta y^2 - \Delta z^2}$ between events (Maudlin, 1993, p. 196). By determining Δs for any pair of events, "the particle trajectories can be embedded in a Minkowski spacetime up to a global space and time translation, rigid rotations, Lorentz transformation, parity, and time inversion" (Maudlin, 1993, pp. 196–197) – just as the relationalist wants to have it. "Once so embedded, the relationist can make use of the full mathematical resources of substantivalist spacetime" (Maudlin, 1993, p. 197). In particular, Minkowski relationalism is not affected by bucket-type arguments: "the special relativistic spatiotemporal relations between points making up a nonrotating bucket are different from those among the points in a rotating bucket" (Maudlin, 1993, p. 197). This is the same for Newtonian relationalism, but, Maudlin argues, while Newtonian relationalism comes across as rather artificial, Minkowski relationalism "is the only plausible form" (Maudlin, 1993, p. 197) of relationalism with respect to special relativity; or, as Pooley puts it, whilst the relationalist had to be rather "creative" (Pooley, 2013, p. 553) in Newtonian theory, special relativity can be approached "flat-footed" (Pooley, 2013, p. 553). Furthermore, Minkowski relationalism (as Newtonian relationalism) is not affected by the static shift argument – notably, fundamentalism about Minkowski spacetime is, though. For more details, see Maudlin (1993).

17 Pooley (2013, pp. 544–574) carefully distinguishes three general strategies of how to be a relationalist.

18 Maudlin (1993) arrives at the same conclusion, but by employing more subtle argumentation. First, he argues, one might think that adopting plenism is equivalent to adopting fundamentalism right away. Conceptually, however,

the following difference can be noted: "[t]he substantivalist's plenum is one which the fields occupy, but the relationist's is one which the fields constitute" (Maudlin, 1993, p. 200) – just think of a static shift scenario. In this conceptual sense, spacetime fundamentalism and relationalism remain distinct positions. According to Maudlin, there is a crucial caveat, though (cf. Earman & Norton, 1987; for the sake of brevity, I shall leave out these additional subtleties). The translational (rotational) static shift argument is only applicable in homogeneous (isotropic) spacetimes – which is naturally the case in Newtonian theory and special relativity. In general relativity, however, only a subset of the solutions to the Einstein field equations is homogeneous (isotropic) – a homogeneous (isotropic) matter or field distribution is a necessary condition (recall that the spacetime geometry depends on the matter or field distribution in general relativity). Hence, Maudlin argues, the static shift argument, though applicable to the subset, is ineffective: the substantivalist and the relationalist agree that all shifted worlds are ontologically identical (disagreement depends solely on how individuation is understood); in its plenist version, the dispute between substantivalism and relationalism is ultimately "purely verbal" (Maudlin, 1993, p. 201). Maudlin then goes on to argue that the most natural candidate for the relationalist's plenum is g and concludes that this essentially amounts to adopting substantivalism.

19 This section is based on work previously published in Salimkhani (2018). The text has been substantially revised and is reproduced with permission from Springer Nature.

20 According to Di Casola et al. (2015), this is not the most general formulation of the weak equivalence principle, but closest to what they dub Newton's equivalence principle.

21 For a list of empirical tests of the equivalence principle see Everitt et al. (2003). While the precision of Galilei's experiments around 1638 were still quite poor, Eötvös's experiments roughly 250 years later increased the precision considerably.

22 According to Di Casola et al. (2015), this is the weak equivalence principle, which is not equivalent to what they dub Newton's equivalence principle (see above).

23 This section is based on work previously published in Salimkhani (2018). The text has been substantially revised and is reproduced with permission from Springer Nature.

24 Formalistically, this may be supported further by the so-called tetrad or vierbein formalism which offers a coordinate-independent, gauge-theoretic formulation of classical general relativity (e.g., Carroll, 2004, pp. 483–494). Essentially, this is based on the freedom of choosing arbitrary basis vectors $\hat{e}_{(a)}$ at every point of the manifold, such that $g(\hat{e}_{(a)}, \hat{e}_{(b)}) = \eta_{ab}$, rather than using the natural coordinate-dependent basis of partial derivatives $\hat{e}_{(\mu)} = \partial_\mu$ for the tangent space and $\hat{\xi}^{(\mu)} = \hat{d}^\mu$ for the co-tangent space, respectively. While there is a sense in which this formalism makes the relationship between general relativity and the gauge theories of particle physics "much more transparent" (Carroll, 2004, p. 483), important differences remain (regarding the fact that in particle physics we are concerned with the *internal spaces* by which the base manifold is *amended* to form a fibre bundle). Hence, to paraphrase Carroll, one should not get carried away – even the merely formal similarities are limited. Nevertheless, there are attempts to utilise this approach for both getting a better grip on some of the conceptual intricacies of general relativity (Ivanenko &

Sardanashvily, 1983) and giving new impulses for a theory of quantum gravity (Hehl et al., 1976). Typically, such attempts explicitly aim at extending standard general relativity. In particular, the possibility of a non-vanishing torsion tensor is argued to be of relevance (see Hehl, 2007; Hehl et al., 1976) – others dispute this (see Carroll, 2004; Carroll & Field, 1994; and Weinberg's response in Hehl, 2007). But it is certainly of philosophical relevance (with respect to the issue of underdetermination; see Lyre & Eynck, 2003) that some of these theories are empirically equivalent to general relativity; notably, there is *teleparallel gravity*, for example, where the Levi-Civita connection with curvature and without torsion is replaced by a Weitzenböck connection with torsion and without curvature (see Lyre & Eynck, 2003; Knox, 2011).

25 See Reichenbach (1928, pp. 353–354): "die geometrische Deutung der Gravitation ist nur das anschauliche Gewand, in welches die Tatsachenbehauptung gekleidet wird. Man darf nicht das Gewand für den Körper halten, der darunter steckt; sondern man darf nur aus der bestimmten Form des Gewandes auf die Form des Körpers schließen, der es trägt. Aber erst dieser Körper ist das Objekt der Physik". Translation in Giovanelli (2021, p. 243) due to the *Archives of Scientific Philosophy (1891–1953). The Hans Reichenbach Papers. 1891–1953*: "The geometrical interpretation of gravitation is merely the visual cloak in which the factual assertion is dressed. It would be a mistake to confuse the cloak with the body it covers; rather, we may infer the shape of the body from the shape of the cloak it wears. After all, only the body is the object of interest in physics".

26 This section includes work that has previously been published in Salimkhani (2020). The text has been substantially revised and extended and is reprinted with permission from Elsevier.

27 Electrons, quarks, photons, and gluons are typically conceived of as excitations of underlying fields. Material bodies are built from electrons and quarks and their electromagnetic and strong interactions (mediated by photons and gluons, respectively). It is important to note that charged particles or fields do not generally exhibit geodesic motion (unless, of course, the charge is rendered ineffective, for example, because the corresponding field, say, an electromagnetic field, is not present); see Menon et al. (2020) for examining the case of photons, which do not carry charge.

28 Notably, this part seems missing – at least, to the best of my knowledge – in the second edition of the book. The general rationale, however, remains the same – as formulations like the following show: "only worldlines that are not geodesical need a causal explanation for their taking the shape in spacetime that they have. The geodesical world lines are as they are because of the projective structure of spacetime. ... let a tangent vector simply 'follow its own nose' through the projective manifold. This is what a particle always does unless caused to do otherwise. That explains why cause-free particles do what they do. There can be no causal explanation why force-free worldlines are geodesical. That springs direct from spacetime structure – from its shape in a general sense" (Nerlich, 1994, p. 231).

29 The verdict that test particles do not "read" (Brown, 2005, p. 24) the spacetime geometry is, according to Brown, supported by the fact that geodesic motion is only approximate.

30 Notably, DiSalle (1995, p. 327) seems to interpret it similarly (see Brown, 2005, p. 25): "When we say that a free particle follows, while a particle experiencing a force deviates from, a geodesic of spacetime, ... we are giving the physical definition of a spacetime geodesic. To say that spacetime has the affine

structure ... is not to postulate some hidden entity to explain the appearances, but rather to say that empirical facts support a system of physical laws that incorporates such a definition" (DiSalle, 1995, p. 327).

31 Influential critical accounts of the dynamical approach include Janssen (2002a; 2002b), Acuña (2012), and Weatherall (2021).

32 Read (2019) interprets this to mean that chronogeometricity is a property that metrical structures "do not have ... essentially", viz., "in all solutions of any theory in which they appear" (Read, 2019, p. 104).

33 Note that, in principle, the geometrical approach can view chronogeometricity as an either non-fundamental or fundamental intrinsic property of the metric (namely, as derivative on other properties of the metric, e.g., its symmetry properties, or not).

34 Anticipating this, DiSalle argues that (see also Brown, 2005, p. 142): "space-time theories do not claim that some unobservable thing is the cause of observable effects. Instead they make a more restricted, but perhaps more profound and certainly more useful claim: that particular physical processes, governed by ... physical laws, can be represented by aspects of geometrical structure .... And this claim provides the only physically meaningful sense in which the universe can be said to have a geometrical structure. More precisely, a claim such as Minkowski's ... that the structure determined by a particular set of physical laws is the structure of a particular four-dimensional space, expresses the only physically meaningful sense in which the universe can be said to have a space-time structure" (DiSalle, 1995, p. 333).

35 Read (2020) raises doubts that Read et al. (2018) are correct in taking the strong equivalence principle as a necessary condition for chronogeometricity. He argues that there may be other ways to gain "operational access to the metric field" (Read, 2020, p. 181) – explored by Ehlers et al. (1972).

36 Read (2019) himself interprets GA-GR as claiming that chronogeometricity is a property which metrical structures "*do* have ... essentially", viz., "in all solutions of any theory in which they appear" (Read, 2019, p. 104).

37 Of course, the matter fields may have additional dynamical symmetry properties that are not fixed by the strong equivalence principle. See also Salimkhani (2022).

38 Instead of "non-gravitational fields", I shall use "matter fields" here. This is to anticipate the newly introduced gravitational "matter" field h in the spin-2 approach in Chapter 4.

39 This section is based on a joint paper with Niels Linnemann. See Linnemann and Salimkhani (2021). The text has been revised and slightly extended; it is included with the permission of Niels Linnemann.

40 The term "constructive" was coined by Brown himself (Brown, 1993, 2005, p. 132), motivated by the distinction between principle and constructive theories (Brown, 2005, pp. 71–73). However, as an important side note, Brown's use of the term deviates from Norton's and my own. This becomes most striking when considering that, on Brown's view, there is a sense in which substantivalism about spacetime is constructive (because it gives fundamental, rather than effective explanations). I do not use Brown's sense of the term here.

41 This is to say that the mathematical representation of physical fields requires recourse to a manifold background structure: they are defined as maps from a manifold onto respective target spaces.

42 Let us add to the standard presentation of the dynamical approach proponent that the dynamical approach arguably also reduces the g field's *causal* property – which is a core spatiotemporal property – to the mutual interactions of the g field

and the matter fields. The causal property of g is its property that matter fields and test particles can only move in accordance with its conformal structure.

43 Pooley argues that "the advocate of the dynamical approach need not be understood as eschewing all primitive spatiotemporal notions (*pace* Norton 2008)" (Pooley, 2013, p. 572), since "the project was to reduce chronogeometric facts to symmetries, not to recover the entire spatiotemporal nature of the world from no spatiotemporal assumptions whatsoever" (Pooley, 2013, p. 573); "[i]n particular, one might take as basic the 'topological' extendedness of the material world in four dimensions" (Pooley, 2013, p. 572). Pooley then indicates that this understanding of the standard dynamical approach might be thought of as akin to Huggett's (2006) regularity relationalism. We agree with Pooley that one has to acknowledge that the 'half-way' approach is a legitimate and ontologically novel position that differs from the standard fundamentalist conception, even though it is not full constructivist.

44 It is worth pointing out that, recently, Binkoski (2019) has identified this issue of "geometric coordination" as a problem for any programme that attempts to render spacetime as a structural quality of each field separately.

45 On the contrary, Norton seems to maintain that Brown's two miracles originate from the "third miracle": "That they agree is attributed to what Brown labels two 'miracles'. One is the 'conspiracy' ... that all free particles ... indicate the same set of inertial motions. The 'second great miracle' ... is that the changes of length of moving rods and dilations of moving clocks are always the same, no matter what their material constitutions. These miracles are really just oblique ways to state the existence of a single preferred set of standard coordinate systems (x, y, z, t) that transcends any particular matter theory" (Norton, 2008, p. 829).

46 A more general account builds on that of Einstein algebras à la Geroch. See Bain (2003) for a philosophical introduction.

47 Recall the debate between, on the one hand, the different priority readings by, for example, Brown and Pooley (2001), as proponents of the dynamical approach, and Balashov and Janssen (2003), as proponents of the geometrical approach, and, on the other hand, a joint reading according to Acuña (2012).

48 We have recently become aware of a criticism of Menon (2019) by Chen and Fritz (2021) to the effect that his approach privileges scalar fields in terms of which all other kind of fields have to be expressed. Superficially, this seems to complement our sentiment that the algebraic structure is just the manifold in disguise: their observation entails that the algebraic structure linked to scalar fields is some background structure for (the other) fields. However, we do insist that merely re-representing other fields algebraically – as Chen and Fritz aspire to do – suffers from the same issue just described for Menon's scalar model. Accordingly, we view Chen and Fritz (2021) as a genuine continuation of Menon's proposal.

49 Notably, this move has be seen as rather unsatisfying or even as a dubious sleight of hand (e.g., Belot, 2000).

50 We take it to be analytic to the Humean/regularity relationist understanding of supervenience basis that what is conceived as supervenience basis is ontologically independent of the objects and patterns that are supposed to be described through it. Of course, that might be questioned but then the whole account of Stevens looks questionable to begin with.

51 Note that the commitment to essentialism here is relatively modest: one only commits to a minimal set of properties anyone with essentialist intuitions

towards space and time would want to agree on (or at least, there do not seem to be relevant positions that are thereby excluded). See Linnemann (2021) for a longer defence along these lines.

52 Norton comments on (different, but sufficiently similar) remarks in Brown (2005, pp. 134–136) that he "share[s] his [Brown's] aversion to unnecessary ontology, but I also fear the opposite extreme of excessive skepticism" (Norton, 2008, p. 831). For Norton, "the common-origin inferences ... provide good reasons to reify" (Norton, 2008, p. 831).

53 One way of making the notion of "totality of interaction events" formally explicit is presented by Westman and Sonego (2009) and taken up by Hardy (2016, Section 9.4): The basic idea on the Westman–Sonego approach is to form a space of local beables spanned by values corresponding to scalar contractions of the fields; each point of the space is a physical event.

54 One may spell out the "needles only" perspective for general relativity further by appeal to results on the causal structure in general relativity from Malament (1977) and, more concretely, a potential successor theory of general relativity based on these results, namely causal set theory (see Wüthrich, 2012 for an introduction): causal set theory posits a set of elementary events (in some sense corresponding to the "needle" events), which combined appropriately yield general-relativistic spacetime.

55 That this is not already implied can be seen as follows: whether something exists or not is not a question of what is needed conceptually. Something might not be needed conceptually but still exist (e.g., derivative objects). Also, something may be needed conceptually but nevertheless not exist *physically* (e.g., numbers). This also makes for a general scepticism regarding whether indispensability arguments, i.e., arguments regarding the *relational* structure, are actually suitable to address issues of realism. One may want to free realism of relational structure and rather conceptualise realism as the question of whether certain *intrinsic* criteria are met (e.g., having energy for what has *physical* reality).

56 Notably, Menon's algebraic formulation could save Curiel's proposal, as it outlines a way of how to formulate the theory without using the manifold explicitly.

References

Abbott, B. P. et al. (2016). Observation of gravitational waves from a binary black hole merger. *Physical Review Letters, 116(6)*, 061102. doi:10.1103/PhysRevLett.116.061102

Abbott, B. P. et al. (2017). Gravitational waves and gamma-rays from a binary neutron star merger: GW170817 and GRB 170817A. *The Astrophysical Journal Letters, 848(2)*, L13. doi:10.3847/2041-8213/aa920c

Acuña, P. (2012). Minkowski spacetime and Lorentz invariance: The cart and the horse or two sides of a single coin? *Studies in History and Philosophy of Science Part B: Studies in History and Philosophy of Modern Physics, 55*, 1–12. doi:10.1016/j.shpsb.2016.04.002

Bain, J. (2003). Einstein algebras and the hole argument. *Philosophy of Science, 70(5)*, 1073–1085. doi:10.1086/377390

Balashov, Y., & Janssen, M. (2003). Presentism and relativity. *The British Journal for the Philosophy of Science, 54(2)*, 327–346. doi:10.1093/bjps/54.2.327

Bartels, A. (2013). Why metrical properties are not powers. *Synthese, 190,* 2001–2013. doi:10.1007/s11229-011-9951-3

Bartels, A. (2015). *Naturgesetze in einer kausalen Welt.* Münster: Mentis.

Belot, G. (2000). Geometry and motion. *The British Journal for the Philosophy of Science, 51,* 561–595. doi:10.1093/bjps/51.4.561

Binkoski, J. (2019). Geometry, fields, and spacetime. *The British Journal for the Philosophy of Science, 70(4),* 1097–1117. doi:10.1093/bjps/axy002

Boran, S., Desai, S., Kahya, E. O., & Woodard, R. P. (2018). GW170817 falsifies dark matter emulators. *Physical Review D, 97(4),* 041501. doi:10.1103/PhysRevD.97.041501

Brown, H. R. (1993). Correspondence, invariance and heuristics in the emergence of special relativity. In S. French, & H. Kamminga (Eds.), *Correspondence, Invariance and Heuristics. Essays in Honour of Heinz Post. Boston Studies in the Philosophy of Science,* Vol. 148 (pp. 256–272). Dordrecht: Springer.

Brown, H. R. (1997). On the role of special relativity in general relativity. *International Studies in the Philosophy of Science, 11(1),* 67–81. doi:10.1080/02698599708573551

Brown, H. R. (2005). *Physical Relativity: Space-Time Structure from a Dynamical Perspective.* Oxford: Oxford University Press.

Brown, H. R., & Pooley, O. (2001). The Origins of the Spacetime Metric: Bell's Lorentzian Pedagogy and Its Significance in General Relativity. In C. Callender, & N. Huggett (Eds.), *Physics Meets Philosophy at the Planck Scale* (pp. 256–272). Cambridge: Cambridge University Press.

Brown, H. R., & Pooley, O. (2006). Minkowski Space-Time: A Glorious Non-Entity. In D. Dieks (Ed.), *The Ontology of Spacetime. Philosophy and Foundations of Physics, Vol. 1* (pp. 67–89). Amsterdam: Elsevier.

Brown, H. R., & Read, J. (2016). Clarifying possible misconceptions in the foundations of general relativity. *American Journal of Physics, 84,* 327–334. doi:10.1119/1.4943264

Brown, H. R., & Read, J. (2021). The Dynamical Approach to Spacetime Theories. In E. K. Wilson (Ed.), *The Routledge Companion to Philosophy of Physics* (pp. 70–85). London/New York, NY: Routledge.

Butterfield, J. (2007). Reconsidering relativistic causality. *International Studies in the Philosophy of Science, 21(3),* 295–328. doi:10.1080/02698590701589585

Carroll, S. (2004). *Spacetime and Geometry: An Introduction to General Relativity.* San Francisco, CA: Addison Wesley.

Carroll, S., & Field, G. (1994). Consequences of propagating torsion in connection dynamic theories of gravity. *Physical Review D, 50(6),* 3867. doi:10.1103/physrevd.50.3867

Chen, L., & Fritz, T. (2021). An algebraic approach to physical fields. *Studies in History and Philosophy of Science, 89,* 188–201. doi:10.1016/j.shpsa.2021.08.011

Curiel, E. (2017). A Primer on Energy Conditions. In D. Lehmkuhl, G. Schiemann, & E. Scholz (Eds.), *Towards a Theory of Spacetime Theories. Einstein Studies, Vol. 13* (pp. 43–104). New, York, NY: Birkhäuser. doi:10.1007/978-1-4939-3210-8_3

Curiel, E. (2018). On the existence of spacetime structure. *The British Journal for the Philosophy of Science, 69(2),* 447–483. doi:10.1093/bjps/axw014

Di Casola, E., Liberati, S., & Sonego, S. (2015). Nonequivalence of equivalence principles. *American Journal of Physics, 83(1)*, 39–46. doi:10.1119/1.4895342

DiSalle, R. (1995). Spacetime theory as physical geometry. *Erkenntnis, 42*, 317–337. doi:10.1007/BF01129008

Dürr, P. M. (2019). It ain't necessarily so: Gravitational waves and energy transport. *Studies in History and Philosophy of Science Part B: Studies in History and Philosophy of Modern Physics, 65*, 25–40. doi:10.1016/j.shpsb.2018.08.005

Earman, J. (1989). *World Enough and Space-Time*. Cambridge, MA: MIT Press.

Earman, J., & Norton, J. D. (1987). What price spacetime substantivalism? The hole story. *The British Journal for the Philosophy of Science, 38(4)*, 515–525. doi:10.1093/bjps/38.4.515

Ehlers, J., Pirani, F. A., & Schild, A. (1972). The Geometry of Free Fall and Light Propagation. In L. O'Reifeartaigh (Ed.), *General Relativity: Papers in Honour of J. L. Synge* (pp. 63–84). Oxford: Clarendon Press.

Einstein, A. (1905). Zur Elektrodynamik bewegter Körper. *Annalen der Physik, 322(10)*, 891–921. doi:10.1002/andp.19053221004

Einstein, A. (1923). On the Electrodynamics of Moving Bodies. In H. A. Lorentz, A. Einstein, H. Minkowski, & H. Weyl (Eds.), *The Principle of Relativity* (pp. 35–65). New York, NY: Dover. Reprint. Translated by W. Perrett and G. B. Jeffery. Translated from "Zur Elektrodynamik bewegter Körper", Annalen der Physik, 17, 1905.

Einstein, A. (2015). *Relativity. The Special and the General Theory. 100th Anniversary Edition*. Princeton, NJ: Princeton University Press.

Everitt, C., Damour, T., Nordtvedt, K., & Reinhard, R. (2003). Historical perspective on testing the equivalence principle. *Advances in Space Research*, 1297–1300.

Fletcher, S. (2020). Approximate Local Poincaré Spacetime Symmetry in General Relativity. In C. Beisbart, T. Sauer, & C. Wüthrich (Eds.), *Thinking About Space and Time. Einstein Studies, Vol 15* (pp. 247–267). Cham: Birkhäuser. doi:10.1007/978-3-030-47782-0_12

Friedman, M. (1983). *Foundations of Space-Time Theories: Relativistic Physics and Philosophy of Science*. Princeton, NJ: Princeton University Press.

Geroch, R. (1972). Einstein algebras. *Communications in Mathematical Physics, 26(4)*, 271–275. doi:10.1007/BF01645521

Giovanelli, M. (2021). 'Geometrization of Physics' vs. 'Physicalization of Geometry'. The Untranslated Appendix to Reichenbach's Philosophie der Raum-Zeit-Lehre. In S. Lutz, & A. T. Tuboly, *Logical Empiricism and the Physical Sciences. From Philosophy of Nature to Physics* (pp. 224–261).

Hardy, L. (2016). Operational general relativity: Possibilistic, probabilistic, and quantum. https://arxiv.org/abs/1608.06940

Hehl, F. W. (2007). Note on the torsion tensor. *Physics Today, 60(3)*, 16. doi:10.1063/1.2718743

Hehl, F. W., Von Der Heyde, P., Kerlick, G. D., & Nester, J. M. (1976). General relativity with spin and torsion: Foundations and prospects. *Reviews of Modern Physics, 48(3)*, 393–416. doi:10.1103/RevModPhys.48.393

Hoefer, C. (1996). The metaphysics of space-time substantivalism. *The Journal of Philosophy, 93(1)*, 5–27. doi:10.2307/2941016

Huggett, N. (2006). The regularity account of relational spacetime. *Mind, 115(457)*, 41–73. doi:10.1093/mind/fzl041

Huggett, N., Lizzi, F., & Menon, T. (2021). Missing the point in noncommutative geometry. *Synthese, 199*, 4695–4728. doi:10.1007/s11229-020-02998-1

Huggett, N., & Wüthrich, C. (book manuscript). *Out of Nowhere.*

Ivanenko, D., & Sardanashvily, G. (1983). The Gauge treatment of gravity. *Physics Reports, 94(1)*, 1–45. doi:10.1016/0370-1573(83)90046-7

Janssen, M. (2002a). COI stories: Explanation and evidence in the history of science. *Perspectives on Science, 10(4)*, 457–522. doi:10.1162/106361402322288066

Janssen, M. (2002b). Reconsidering a scientific revolution: The case of Einstein versus Lorentz. *Physics in Perspective, 4*(4), 421–446.

Knox, E. (2011). Newton–Cartan theory and teleparallel gravity: The force of a formulation. *Studies in History and Philosophy of Science Part B: Studies in History and Philosophy of Modern Physics, 42(4)*, 264–275. doi:10.1016/j.shpsb.2011.09.003

Knox, E. (2019). Physical relativity from a functionalist perspective. *Studies in History and Philosophy of Science Part B: Studies in History and Philosophy of Modern Physics, 67*, 118–124. doi:10.1016/j.shpsb.2017.09.008

Lehmkuhl, D. (2008). Is Spacetime a Gravitational Field? In D. Dieks (Ed.), *The Ontology of Spacetime Vol. II. Philosophy and Foundations of Physics, Vol. 4* (pp. 83–110). Amsterdam: Elsevier.

Lehmkuhl, D. (2014). Why Einstein did not believe that general relativity geometrizes gravity. *Studies in History and Philosophy of Science Part B: Studies in History and Philosophy of Modern Physics, 46*(Part B), 316–326. doi:10.1016/j.shpsb.2013.08.002

Lehmkuhl, D. (2021). The Equivalence Principle(s). In E. Knox, & A. Wilson (Eds.), *The Routledge Companion to Philosophy of Physics* (pp. 125–144). New York, NY/London: Routledge.

Linnemann, N. (2021). On the empirical coherence and the spatiotemporal gap problem in quantum gravity: And why functionalism does not (have to) help. *Synthese, 199*(Suppl 2), 395–412. doi:10.1007/s11229-020-02659-3

Linnemann, N., & Salimkhani, K. (2021). The Constructivist's Programme and the Problem of Pregeometry. http://philsci-archive.pitt.edu/20035/

Lyre, H., & Eynck, T. O. (2003). Curve it, Gauge it, or leave it? *Journal for General Philosophy of Science, 34(2)*, 277–303. doi:10.1023/B:JGPS.0000005161.79937.ab

Maggiore, M. (2008). *Gravitational Waves. Volume 1: Theory and Experiment.* Oxford: Oxford University Press.

Malament, D. B. (1977). The class of continuous timelike curves determines the topology of spacetime. *Journal of Mathematical Physics, 18(7)*, 1399–1404. doi:10.1063/1.523436

Malament, D. B. (2012). A Remark About the "Geodesic Principle" in General Relativity. In M. Frappier, D. H. Brown, & R. DiSalle (Eds.), *Analysis and Interpretation in the Exact Sciences. Essays in Honour of William Demopoulos* (pp. 245–252). Dordrecht: Springer.

Martens, N. C. (2019). The metaphysics of emergent spacetime theories. *Philosophy Compass, 14(7)*, e12596. Doi:10.1111/phc3.12596

Maudlin, T. (1993). Buckets of water and waves of space: Why spacetime is probably a substance. *Philosophy of Science, 60(2)*, 183–203. doi:10.1086/289728

Maudlin, T. (1996). On the unification of physics. *The Journal of Philosophy, 93(3)*, 129–144. doi:10.2307/2940873

Maudlin, T. (2012). *Philosophy of Physics: Space and Time*. Princeton, NJ: Princeton University Press.

Menon, T. (2019). Algebraic fields and the dynamical approach to physical geometry. *Philosophy of Science, 86(5)*, 1273–1283. doi:10.1086/705508

Menon, T., Linnemann, N., & Read, J. (2020). Clocks and chronogeometry: Rotating spacetimes and the relativistic null hypothesis. *The British Journal for the Philosophy of Science, 71(4)*, 1287–1317. doi:10.1093/bjps/axy055

Minkowski, H. (1909). Raum und Zeit. *Physikalische Zeitschrift, 10*, 104–111.

Minkowski, H. (1923). Space and Time. In H. A. Lorentz, A. Einstein, H. Minkowski, & H. Weyl (Eds.), *The Principle of Relativity* (pp. 73–91). New York, NY: Dover. Reprint. Translated by W. Perrett and G. B. Jeffery. Translation of an Address delivered at the 80th Assembly of German Natural Scientists and Physicians, at Cologne, 21 December, 1908.

Misner, C. W., Thorne, K. S., & Wheeler, J. A. (1973). *Gravitation*. Princeton, NJ: Princeton University Press.

Nerlich, G. (1976). *The Shape of Space*. Cambridge: Cambridge University Press.

Nerlich, G. (1994). *The Shape of Space* (2nd ed.). Cambridge: Cambridge University Press.

Norton, J. D. (2008). Why constructive relativity fails. *The British Journal for the Philosophy of Science, 59(4)*, 821–834. doi:10.1093/bjps/axn046

Norton, J. D. (2019). The Hole Argument. In E. N. Zalta (Ed.), *The Stanford Encyclopedia of Philosophy* (Summer 2019 ed.). Metaphysics Research Lab, Stanford University. https://plato.stanford.edu/archives/sum2019/entries/spacetime-holearg/

Pooley, O. (2013). Substantivalist and Relationalist Approaches to Spacetime. In R. Batterman (Ed.), *The Oxford Handbook of Philosophy of Physics* (pp. 522–586). Oxford: Oxford University Press.

Pooley, O. (2017). Background Independence, Diffeomorphism Invariance, and the Meaning of Coordinates. In D. Lehmkuhl, G. Schiemann, & E. Scholz (Eds.), *Towards a Theory of Spacetime Theories. Einstein Studies, Vol. 13* (pp. 105–143). New York, NY: Birkhäuser. doi:10.1007/978-1-4939-3210-8_4

Read, J. (2018). Functional gravitational energy. *The British Journal for the Philosophy of Science, 71(1)*, 205–232. doi:10.1093/bjps/axx048

Read, J. (2019). On miracles and spacetime. *Studies in History and Philosophy of Science Part B: Studies in History and Philosophy of Modern Physics, 65*, 103–111. doi:10.1016/j.shpsb.2018.10.002

Read, J. (2020). Explanation, Geometry, and Conspiracy in Relativity Theory. In C. Beisbart, T. Sauer, & C. Wüthrich (Eds.), *Thinking About Space and Time: 100 Years of Applying and Interpreting General Relativity. Einstein Studies, Vol. 15* (pp. 173–205). Basel: Birkhäuser.

Read, J., Brown, H. R., & Lehmkuhl, D. (2018). Two miracles of general relativity. *Studies in History and Philosophy of Science Part B: Studies in History and Philosophy of Modern Physics, 64*, 14–25. doi:10.1016/j.shpsb.2018.03.001

Reichenbach, H. (1928). *Philosophie der Raum-Zeit-Lehre*. Berlin/Leipzig: Walter de Gruyter. Reprint in A. Kamlah and M. Reichenbach (Eds.), *Reichenbach, Gesammelte Werke in 9 Bänden*. Braunschweig/Wiesbaden: Vieweg, 1977.

Reichenbach, H. (1958). *Philosophy of Space and Time*. New York, NY: Dover Publications.

Rosen, N. (1940a). General relativity and flat space. I. *Physical Review*, *57*(2), 147–150. doi:10.1103/PhysRev.57.147

Rosen, N. (1940b). General relativity and flat space. II. *Physical Review*, *57(2)*, 150–153. doi:10.1103/PhysRev.57.150

Rosenstock, S., Barrett, T. W., & Weatherall, J. O. (2015). On Einstein Algebras and relativistic spacetimes. *Studies in History and Philosophy of Science Part B: Studies in History and Philosophy of Modern Physics*, *52*, 309–316. doi:10.1016/j.shpsb.2015.09.003

Rynasiewicz, R. (1992). Rings, holes and substantivalism: On the program of Leibniz Algebras. *Philosophy of Science*, *59(4)*, 572–589. doi:10.1086/289696

Salimkhani, K. (2018). Quantum Gravity: A Dogma of Unification? In A. Christian, D. Hommen, N. Retzlaff, & G. Schurz (Eds.), *Philosophy of Science: Between the Natural Sciences, the Social Sciences, and the Humanities* (pp. 23–41). Cham: Springer International Publishing. doi:10.1007/978-3-319-72577-2_2

Salimkhani, K. (2020). The dynamical approach to spin-2 gravity. *Studies in History and Philosophy of Science Part B: Studies in History and Philosophy of Modern Physics*, *72*, 29–45. doi:10.1016/j.shpsb.2020.05.002

Salimkhani, K. (2022). A Dynamical Perspective on the Arrow of Time. http://philsci-archive.pitt.edu/20852/

Schrödinger, E. (1950). *Space-Time Structure*. Cambridge: Cambridge at the University Press.

Stachel, J. (1993). Development of the Concepts of Space, Time and Space-Time from Newton to Einstein. In A. Ashtekar (Ed.), *100 Years of Relativity. Space-Time Structure: Einstein and Beyond* (pp. 3–36). London: World Scientific.

Stevens, S. (2018). Regularity relationalism and the constructivist project. *The British Journal for the Philosophy of Science*, *71(1)*, 353–372. doi:10.1093/bjps/axx037

Wald, R. M. (1984). *General Relativity*. Chicago, IL: The University of Chicago Press.

Weatherall, J. O. (2021). Two dogmas of dynamicism. *Synthese*, *199*(Suppl 2), 253–275. doi:10.1007/s11229-020-02880-0

Weinberg, S. (1972). *Gravitation and Cosmology: Principles and Applications of the General Theory of Relativity*. New York, NY: Wiley.

Westman, H., & Sonego, S. (2009). Coordinates, observables and symmetry in relativity. *Annals of Physics*, *324*(8), 1585–1611.

Weyl, H. (1970). *Raum, Zeit, Materie* (6th ed.). Berlin/Heidelberg: Springer-Verlag.

Wüthrich, C. (2012). In Search of Lost Spacetime: Philosophical Issues Arising in Quantum Gravity. English Version of 'A la recherche de l'espacetemps perdu: questions philosophiques concernant la gravité quantique'. In S. Le Bihan (Ed.), *La Philosophie de la Physique: D'aujourd'hui à demain* (pp. 222–241). Paris: Vuibert, 2013. arXiv:1207.1489 [physics.hist-ph]

4 Classical "Spin-2" Gravity[1]

To study further whether general-relativistic spacetime is fundamental, I will now introduce an alternative formulation of general relativity: classical "spin-2" gravity. The features of the classical and the quantum version (see Section 5.2) of the spin-2 approach as well as their relations to Einstein's theory of general relativity have been worked out by a number of physicists[2] – for a concise review see Preskill and Thorne (1995), for an extensive list of references see also Pitts (2016). Amongst others, the spin-2 approach demonstrates that Einstein's field equations can be derived "non-geometrically" (Deser, 1970, p. 9) – i.e., without drawing on geometrical notions, but by employing the standard field-theoretical techniques.

After presenting and discussing the physics of the classical spin-2 theory of gravity, I draw some philosophical lessons for the debate on the dynamical approach and, subsequently, for the question of whether spacetime is fundamental. In particular, I argue that there is a sense in which (aspects of) spacetime can indeed be considered non-fundamental in a dynamical spin-2 reading of general relativity.

4.1 Reviewing Classical Spin-2 Theory

First, I introduce the potentially surprising notion of a "*classical* spin-2 field" and its specific connection to quantum physics – not with full mathematical rigour, but in some detail; this part may be skipped. Then, I present the classical spin-2 approach to general relativity. After a brief summary of the key philosophical insights in Section 4.1.3, I draw attention to three issues that are relevant for my further philosophical assessment of the classical spin-2 approach in Section 4.1.4.

4.1.1 *Preliminaries*

First of all, what exactly is a "classical spin-2 field"? After all, spin is a quantum property, for all we know. Although this is of course correct, I

DOI: 10.4324/9781003404149-4

shall argue that "classical spin-2" is *not* a misnomer – there is some deep conceptual sense to the notion. Roughly, this is because the spin properties associated with quantum fields express certain rotational symmetry properties that have a specific classical counterpart (see Carroll, 2004; Maggiore, 2008; Preskill & Thorne, 1995; Weinberg, 1972; 1995). Essentially, the concept of spin derives from the more general concept of invariance under external symmetry transformations, so that an analogue to spin can also be classically defined.

Consider a generic classical second-rank tensor $X_{\mu\nu}$ with 16 components[3] (or degrees of freedom). $X_{\mu\nu}$ decomposes into the following classical (i.e., non-unitary) irreducible representations of the Lorentz group[4]: an antisymmetric tensor $A_{\mu\nu} = -A_{\nu\mu}$ with six independent degrees of freedom, a symmetric ($S_{\mu\nu} = S_{\nu\mu}$) and traceless ($S_\mu^\mu = 0$) tensor $S_{\mu\nu}$ with nine independent degrees of freedom, and a (scalar) trace $X \equiv X_\mu^\mu$ (one degree of freedom). For technical reasons one alternatively often considers the symmetric tensor $S'_{\mu\nu} = S_{\mu\nu} + X\,\eta_{\mu\nu}$ with non-vanishing trace and ten independent degrees of freedom; the classical polarisation tensor $e_{\mu\nu}$ that appears in classical plane wave solutions of the form $h_{\mu\nu}(x) = e_{\mu\nu}\,\exp(ik_\lambda x^\lambda) + e^*_{\mu\nu}\,\exp(-ik_\lambda x^\lambda)$, with the position vector x^μ and the wave vector k^μ, is an example for such a tensor $S_{\mu\nu}$ or $S'_{\mu\nu}$.

With respect to the rotation subgroup SO(3), the six degrees of freedom of the antisymmetric tensor decompose into two (generally massive) "spin-1" vectors with three components ($2s + 1$, with $s = 1$); while the symmetric and traceless tensor decomposes into a "spin-0" scalar, a spin-1 vector, and the five components ($2s + 1$, with $s = 2$) of a (generally massive) "spin-2" tensor (Maggiore, 2008, p. 70).

It is important to note that a *massless* spin-2, which yields long-range interactions, only has two physically significant degrees of freedom. Accordingly, there is a mismatch between the number of physical states and the representational degrees of freedom. This is at the heart of why both classical and quantum theories of massless spin-1 and spin-2 fields are "highly constrained" (Preskill & Thorne, 1995, p. xi) – in contrast to scalar theories (Ortín, 2015, p. 47); see also Appendix A.2.

The two physically significant degrees of freedom of a massless spin-2 correspond to its *helicity* eigenvalues $\pm h$. In relativistic quantum particle theory, the helicity of a quantum particle is the projection of the particle's spin onto the particle's direction of momentum. Classically, this corresponds to the following: "a plane wave ψ which is transformed by a rotation of any angle θ about the direction of propagation into $\psi' = e^{ih\theta}\psi$ is said to have helicity h" (Weinberg, 1972, p. 257). Electromagnetic waves can be decomposed into parts with helicity ± 1 and 0, but only the ± 1 helicities are physically significant. Similarly, gravitational waves have physically significant helicities ± 2, and physically insignificant helicities ± 1 and 0.

In quantum theory this corresponds to photons and gravitons being spin-1 and spin-2 particles, respectively.

> This is what we mean when we say, speaking classically, that electro-magnetism and gravitation are carried by waves of spin 1 and spin 2, respectively.
>
> (Weinberg, 1972, p. 258)

The symmetric and traceless tensor represents the polarisation tensor of a massless spin-2 field for appropriate invariance conditions (i.e., gauge invariance) that eliminate the seven spurious degrees of freedom as un-physical (see also Ogievetsky & Polubarinov, 1965; Weinberg, 1972, pp. 255–258). In this sense, the (classical) symmetric and traceless tensor field $S_{\mu\nu}$ corresponding to the massless spin-2 particle is a "classical spin-2 field". It is standardly denoted as h (not to be confused with helicity, but unfor-tunately usually denoted the same way – here in italics, though), or by its components $h_{\mu\nu}$. As already mentioned, for technical reasons, h is often taken to represent a symmetric but not traceless tensor with eight degrees of freedom to be eliminated (see Fang & Frønsdal, 1979, pp. 2264–2265; Maggiore, 2008, p. 71).

Let me dwell on the spin issue a bit further. As we have seen, the quantum-mechanical spin properties have a counterpart in classical physics. Just as quantum spin properties are used to classify quantum particles, classical "spin" properties can be used to classify classical fields – notably, with-out reference to quantum physics. Classical "spin" is not something that we can only construct with quantum mechanics in mind. Rather, it is a direct consequence of a theory's "external" symmetries (as opposed to the "internal" gauge symmetries introduced by modern particle physics). Ex-ternal symmetries, like rotations or transformations, have spatiotemporal significance and already appear in classical physics. In this subtle sense, spin is *not* a purely quantum-mechanical notion.[5] Rather, the quantum-mechanical spin proves to be a special case of a more generic concept that already appears at the level of classical physics.

Both classical fields, like the electromagnetic field, and quantum parti-cles, like the electron, are classified – one may even say: in part defined – by group theoretical considerations (or, more precisely, by representation the-ory). Roughly speaking, we consider the homogeneous Lorentz group for classical fields and the inhomogeneous Lorentz group (or Poincaré group) for quantum particles. A classical field is then described by a specific non-unitary irreducible representation of the homogeneous Lorentz group; likewise, a quantum particle is described by a unitary irreducible repre-sentation of the inhomogeneous Lorentz group. More precisely, we clas-sify the classical fields by the finite-dimensional irreducible non-unitary representations of the homogeneous Lorentz group, and the quantum particle species by the non-negative energy infinite-dimensional irreducible unitary representations of the inhomogeneous Lorentz group (e.g., Tung, 2003).

Now, these irreducible representations are distinguished by only two labels. The labelling is due to the two Casimir invariants of the Lorentz group and the Poincaré group, respectively.[6] A Casimir invariant of a group commutes with all elements of the group, and hence with all elements of any irreducible representation of the group – its eigenvalues can therefore serve to specify and label the different irreducible representations. In particular, one of the Casimir invariants of the Poincaré group and both Casimir invariants of the homogeneous Lorentz group are related to spin.[7]

Hence, the fact that spin can serve as a label to distinguish particle species in quantum theory is a direct consequence of the requirement that the laws of physics be invariant under Lorentz transformations, and is foreshadowed by the fact that classical "spin" can serve as a label to distinguish the field species in classical physics. The standard conviction that spin is "a quantum degree of freedom ... which has no classical counterpart" (Shankar, 2008, p. 373) – based on the observation that, historically, spin has been introduced to physics to describe certain phenomena of (non-relativistic) quantum mechanics – downplays this foundationally significant aspect.

Conceptually speaking, there fundamentally being a spin property is indeed the result of specifically *(special-)relativistic* quantum particle theory (or quantum field theory, for that matter) – and that in three ways: first, the very notion of a quantum particle ultimately derives from quantum field theory, which is essentially built on joining the principles of quantum mechanics and special relativity (e.g., Weinberg, 1999). Second, as presented above, the particular spin property of a relativistic quantum particle reflects the fact that the corresponding relativistic quantum field transforms in accordance with a particular irreducible representation of the underlying external symmetry group. The different fields can therefore be labelled, amongst others, by a number $s = n/2$, with $n \in \mathbb{N}_0$ that is interpreted as the spin. As mentioned, this is a direct consequence of the invariance requirements posed by special relativity, namely Lorentz invariance.[8] Third, the spin-statistics theorem divides relativistic quantum particles into fermions with half-integer spin, and bosons with integer spin.[9] Lorentz invariance or special-relativistic light cone structure[10] is crucial for proving the spin-statistics theorem (see Streater & Wightman, 2000) – which arguably is not surprising given that Lorentz invariance is the very reason for why spin is a defining property of particles.

Since Lorentz invariance is crucial for quantum particles to have spin, classical spin-2 theory is similarly restrictive as its quantum version. The former I present in the following, the latter in Chapter 5.

4.1.2   *Deriving the Einstein Field Equations*

Usually, the classical spin-2 field is introduced in the linearised limit of Einstein's theory of general relativity in terms of the symmetric and traceless tensor $h_{\mu\nu}$ representing a small perturbation on the Minkowski metric $\eta_{\mu\nu}$, i.e., an expansion of $g_{\mu\nu}$ to the first order in $h_{\mu\nu}$ (e.g., Maggiore, 2008):

$$g_{\mu\nu} = \eta_{\mu\nu} + h_{\mu\nu}, \qquad |h_{\mu\nu}| \ll 1. \tag{4.1}$$

This "weak-field" limit of general relativity is convenient for various analyses and applications, especially for weak gravitational waves. However, and this is the point of this section, classical spin-2 theory is distinct from this *ex-post* limit of general relativity. It is a field theory in its own right that provides substantial foundational insights.

To see this, let us forget for a moment about general relativity, its weak field limit, the curved metric g, and the perturbative interpretation of *h*, and consider a classical field theory of a free massless spin-2 field *h* on a fixed[11] flat Minkowski background η.[12] In this context, *h* is treated as any other field defined on flat Minkowski spacetime (e.g., the electromagnetic field). As in any other field theory, *h* is assumed to vanish at infinity,[13] but *not* assumed to be small (see Deser, 1970; 1987; 2010; see also Weinberg, 1972, p. 165). We shall find that this leads us to the Einstein field equations of general relativity and the curved metric g (a few qualifications are briefly discussed in Section 4.1.4). As a result, we may entertain the intriguing idea that this could facilitate an ontological reduction of g to *h* (cf. Sections 4.3 and 5.3).

Historically, several approaches – different in strength – have been pursued.[14] In particular, after some preliminary work on the perturbative expansion of the general-relativistic Lagrangian in flat space,[15] Gupta (1954) first proposed that one may derive the non-linear Einstein field equations from a linear massless spin-2 theory (building on Fierz & Pauli, 1939) by iteratively summing an infinite series of energy-momentum contributions:

> In a nutshell, Gupta's idea is that the nonlinearities of Einstein's theory of the metric field can be understood without founding it on a geometrical interpretation.
>
> (Fang & Frønsdal, 1979, p. 2265)

It is important to note, however, that Gupta – like others who worked out similar accounts (e.g., Feynman) – did not carry out the calculation in full, and, hence, did not properly *prove* that Einstein's theory can be obtained in this way; it is Kraichnan (1955) who first (and independently) obtained Einstein's field equations (see Carlip, 2014; Preskill & Thorne, 1995).

Notably, Deser (1970) then gave the first rigorous and complete derivation that arrives at the full non-linear theory in closed form (i.e., in one step without iteration). In brief, Deser bypasses the infinite series summation (only an additional cubic term appears) by using the Palatini formalism, i.e., by treating the metric and the affine connection independently. Deser's (1970) derivation is also considered the stronger result with respect to presuppositions and scope: in particular, unlike Gupta (1954), Deser (as well as Kraichnan and Feynman, for example) does not simply *assume* that gravity couples to the total energy momentum tensor, but *derives* this from the consistency of the field equations (Preskill & Thorne, 1995).[16]

While the attempts at deriving Einstein's field equations in the spin-2 approach differ with respect to technical details, presuppositions, rigour, and completeness, the reasoning is essentially the same: the full non-linear theory (i.e., general relativity) is argued to follow from the linear free field equations for a massless spin-2 field h, since these equations pose non-trivial consistency conditions that are only[17] satisfied by the full theory.[18]

To be a bit more specific, the derivation of Einstein's field equations can either focus on gauge invariance (e.g., Ogievetsky & Polubarinov, 1965; Sexl & Urbantke, 2002; Thirring, 1959; 1961), or on the mathematical self-consistency of the equations (e.g., Deser, 1970; see Ortín, 2015, p. 89), but ultimately both gauge invariance and self-consistency are part of a full derivation (gauge invariance specifies the linear free field equations to which self-consistency is applied).[19] In the following, I focus on the self-consistency argument that represents the non-linearities in terms of self-interactions of h (Deser, 1970, p. 9). So as to pay particular attention to the aspects relevant to my discussion, I will not give a full derivation of Einstein's field equations here, but only sketch the general rationale.

The linear free field equations for a massless, spin-2 field h, are as follows (see Álvarez, 1989, for more details):

$$\left[\left(\eta_\mu^{\ \alpha}\eta_\nu^{\ \beta}-\eta_{\mu\nu}\eta^{\alpha\beta}\right)\left(\partial_t^2-\partial^i\partial_i\right)+\eta_{\mu\nu}\,\partial^\alpha\partial^\beta+\eta^{\alpha\beta}\,\partial_\mu\partial_\nu-\eta_\mu^{\ \beta}\,\partial^\alpha\partial_\nu-\eta_\nu^{\ \alpha}\,\partial^\beta\partial_\mu\right]h_{\alpha\beta}=0$$

$$(4.2)$$

Deser (1970) does not provide an argument on how to obtain these equations, but seems to rely on the linear limit of the vacuum field equations, i.e., $G_{\mu\nu}^{(L)}(h) \equiv R_{\mu\nu}^{(L)}(h) - \frac{1}{2}R^{(L)}(h)\eta_{\mu\nu} = 0$, where $G_{\mu\nu}^{(L)}(h)$, $R_{\mu\nu}^{(L)}(h)$, and $R^{(L)}(h)$ are the linearised Einstein tensor, the linearised Ricci tensor, and the linearised Ricci scalar for h, respectively. With respect to the philosophical issues, it is important to note that this does not pose a circularity problem: the dependence of the linear free field equations for h on the full non-linear vacuum field equations for g is merely heuristical. In fact, the linear free field equations for h can be obtained independently from general relativity by arguing for the simplest linear spin-2 theory in analogy with electromagnetism (see Álvarez, 1989; Feynman et al., 1995; Fierz & Pauli, 1939; Ortín, 2015). Any "educated guess" regarding the choice of the correct form of the equations (in light of potential alternatives) should essentially be viewed as informed by the demand for empirical adequacy (see also endnote 18).

As mentioned, h is supposed to be a *free* field on η. Accordingly, the equation for h is consistent as it stands; its solutions are the solutions to the wave equation.

Now, we want to go beyond the free theory and introduce interactions between h and other (ordinary) matter fields. Technically, this is done by adding a dynamical source term on the right-hand side of the equation, the (ordinary) matter energy-momentum tensor $T_{\mu\nu}$. Note, however, that the left-hand side of the free field equation is divergenceless. This "divergence identity" of the linear free field equations – which can be identified with the linearised Bianchi identity (Wald, 1986) – poses a severe constraint on any generalisation of the linear free field equations: any modification of the right-hand side must respect this constraint. In particular, a dynamical source in terms of an energy-momentum tensor must have vanishing divergence, i.e., must be conserved; it is interesting to note that a scalar field theory is significantly less restrictive and does not constrain the choice of the energy-momentum tensor (Ortín, 2015, p. 47).

Adding $T_{\mu\nu}$ as a dynamical source on the right-hand side introduces a term with non-vanishing divergence: coupling h to ordinary matter implies – as is derivable from the equations of motion – that the ordinary matter energy-momentum tensor is no longer conserved ($T^{\mu\nu}{}_{,\nu} = \nabla_\nu T^{\mu\nu} \neq 0$).[20] "Naively" coupling h to other matter fields by adding $T_{\mu\nu}$ on the right-hand side renders the equations inconsistent. By coupling h to ordinary matter, energy and momentum can be exchanged between h and the ordinary matter fields, thus, energy-momentum conservation no longer holds for the ordinary matter fields alone.

To do away with this inconsistency, conservation of energy and momentum needs to be restored by including the self-coupling contributions of h to the *total* energy-momentum tensor (i.e., by adding them to the ordinary matter energy-momentum tensor $T_{\mu\nu}$). For the linear theory introduced above in Eq. (4.2), which is obtained from a Lagrangian quadratic in h, this means that we need to include the quadratic contribution ${}^{(2)}\Theta_{\mu\nu}(h)$. However, this makes the formerly linear equations (4.2) *quadratic* in h, which is why we actually need to consider the quadratic equations of motion, i.e., the Lagrangian *cubic* in h, which contributes the *cubic* energy-momentum tensor ${}^{(3)}\Theta_{\mu\nu}(h)$. However, this renders the equations of motion cubic, i.e., the Lagrangian *quartic* in h, and so on.

In this way, an infinite series of contributions to the energy-momentum tensor is forced on us by a mathematical consistency condition (the divergence identity of the linear free field equations for h). Starting from the linear theory, it turns out that it is not sufficient to simply include the self-coupling contributions of that linear theory to restore consistency; only for the full series of self-coupling contributions ($\Theta_{\mu\nu} = \sum_{n=2}^{\infty} {}^{(n)}\Theta_{\mu\nu}(h)$) is energy-momentum conserved, and, hence, the consistency condition satisfied.

We obtain an infinite series of self-coupling contributions of h which can be shown to sum to the full non-linear Einstein equations $G_{\mu\nu}(\eta + h) = -\kappa T_{\mu\nu}$ (Deser, 1970, p. 10). In particular, matter only couples to $\eta_{\mu\nu} + h_{\mu\nu}$ (Carroll, 2004, p. 299) and $\eta_{\mu\nu}$ does not appear in isolation (Kraichnan, 1956, p. 484). Thus, we can re-express $\eta_{\mu\nu} + h_{\mu\nu} \equiv g_{\mu\nu}$. It is crucial to note again that $\eta + h$ is *not* a first-order expansion of g; h is *not* a small perturbation on η but a field in its own right that is then interpreted as the *full* deviation of a (new effective) metric g from η (Deser, 1970; 2010). All non-linear contributions ($\mathcal{O}(h^2)$ and higher), i.e., all self-coupling contributions of h, are absorbed in the now-conserved $T_{\mu\nu}$ on the right-hand side.[21]

Obviously, the notation is supposed to indicate that what is re-expressed as "g" is, due to its featuring in the Einstein field equations, the metric field of general relativity. Looking ahead to certain qualifications regarding the equivalence of full non-linear spin-2 theory and general relativity, one might criticise that, in order to make the potential difference between the two theories transparent, I should rather be talking about g̃ here. This, however, runs the risk of encouraging the following misconception: namely, that the qualifications regarding the equivalence of the two theories would imply that $\tilde{g} \neq g$. They do not. Rather, the *solution spaces* of the two theories are different. Accordingly, $\tilde{g} = g$, but not all general-relativistic g fields can be accommodated in full non-linear spin-2 theory.

To summarise, if *any* matter coupling of h is to be allowed, the self-coupling contributions of h need to be included for consistency reasons. Either the spin-2 theory stays free, or the h field is required to couple with the same strength to *all* fields present (including itself). In other words, it is required to couple to the *total* energy-momentum tensor of the theory (Ortín, 2015, p. 78). Any partial or non-universal coupling to ordinary matter would render the theory inconsistent (see also Kraichnan, 1956). Establishing that h couples universally to all energy and momentum is then interpreted as establishing the strong equivalence principle (see Section 5.2.3) and as suggesting – *ex post* – a geometrical interpretation of the (from the outset *non*-geometrical) theory:

> Consistency has therefore led us to universal coupling, which implies the equivalence principle. It is at this point that the geometrical interpretation of general relativity arises, since all matter now moves in

an effective Riemann space of metric $g^{\mu\nu} \equiv \eta^{\mu\nu} + h^{\mu\nu}$, and so the initial flat 'background' space $\eta^{\mu\nu}$ is no longer observable.

(Deser, 1970, p. 13)

Or, as Kraichnan (1956) puts it with respect to his own derivation:

$\eta_{\mu\lambda}$ appears in the equations of motion only in the combination $g_{\mu\lambda} = \eta_{\mu\lambda} - \lambda f_{\mu\lambda}$. Thus, even though $\eta_{\mu\lambda}$ is explicitly introduced at the outset of the present formulation as the metric of space, it is found to be unobservable and $g_{\mu\lambda}$ is the physically meaningful metric.

(Kraichnan, 1956, p. 484)

I investigate this aspect further in Section 4.2.1.

4.1.3 A First Look at the Philosophical Implications

The most important result is that an interacting classical spin-2 theory is only consistent as the full non-linear spin-2 theory, which is equivalent (up to a few qualifications discussed below) to the standard formulation of general relativity – both feature the g field (either built-in or effectively) and both yield the non-linear Einstein field equations. In the standard set-up, one has (apart from the ordinary matter fields and their dynamics): (a) the generally curved g field and (b) the presumption of the strong equivalence principle – i.e., the fact that locally all laws reduce to the laws of special relativity; in the alternative spin-2 set-up, one has (again, apart from the ordinary matter fields and their dynamics): (a) the flat η field and (b) the h field which, for consistency reasons, couples universally to all energy and momentum including self-coupling.

In the appropriate dynamical reading the *fundamental* posits of these two set-ups are, in the former case, the g field and the strong equivalence principle, and, in the latter case, the new matter field h and its Lorentz-invariant dynamics. The proponent of the dynamical approach interprets the fixed, flat η field as a mere codification of the dynamical symmetry properties of the matter fields.

Looking ahead, the classical spin-2 derivation of general relativity (up to a few qualifications), including the derivation of the universal coupling of h, will come in handy to shed new light on a couple of foundational and philosophical issues in general relativity. In particular, as I shall argue in Section 4.2.1, it is suggested that g is non-fundamental – contrary to the standard fundamentalist view. What is more, with a few qualifications, general relativity can be derived from a consistent special-relativistic spin-2 field theory.

4.1.4   *Three Qualifications*

Before I study the philosophical implications of the spin-2 approach more closely, let me point out a few qualifications regarding the claim that classical spin-2 theory is equivalent to general relativity. First of all, what does "equivalent" mean here? In general, it could mean that (a) the theories are *empirically* equivalent (e.g., regarding their empirical predictions or their accordance with available empirical data); or, that (b) both theories give the same action and field equations (namely, the Einstein-Hilbert action and the Einstein field equations, respectively); or, that (c) both theories have the same solution space. If the two theories are equivalent with respect to their empirical predictions, mathematical structure, and their solution spaces, we can say that the theories are *strictly* equivalent.

Spin-2 theory and general relativity are *not* strictly equivalent. Most importantly, spin-2 theory cannot accommodate all solutions of general relativity, but arguably only globally hyperbolic ones. Still, spin-2 theory and general relativity are equivalent in decisive respects: most importantly, they have the same field equations and are both in accordance with all available empirical data. This will suffice for my purposes. The results of Sections 4.2 and 4.3 do not stand or fall with the equivalence claim. First of all, even if spin-2 theory was not equivalent to general relativity in any sense (which is not true), this obviously would not diminish the foundational and philosophical results for spin-2 theory itself. Second, the two theories are at least empirically equivalent with respect to the well-behaved solutions relevant for describing our universe. Only if this turned out not to be the case, would inferences from full non-linear spin-2 theory to general relativity be blocked – namely, empirically.[22] There is no clear-cut indication so far that classical spin-2 theory is empirically inferior to general relativity. Therefore, all results obtained for spin-2 theory and its effective metric g are transferable to at least the empirically most relevant solutions of general relativity. But to be clear, *prima facie*, some of the following qualifications might call into question whether the results of spin-2 theory can be easily transferred to general relativity. Ultimately, a more comprehensive investigation of these issues is crucial for assessing the robustness of what I propose later for the interpretation of general relativity.

4.1.4.1   *Cosmological Constant Term*

First, the flat background derivation (using a flat metric η defined on a flat manifold) is not able to accommodate a cosmological constant term $\Lambda g_{\mu\nu}$ in the Einstein field equations. This is at odds with the claim that spin-2 theory and general relativity have the same field equations. It might be seen as a minor issue, though, since one could argue that setting the cosmological term to zero is justified because Λ is empirically constrained to be very small (Weinberg, 1972, p. 155). We do, however, know that the cosmological constant is non-zero and that this has important consequences, like the accelerating expansion of the universe (Padmanabhan, 2003). Accordingly, our standard model of cosmology, the ΛCDM model, includes a small, but non-zero cosmological constant.[23] So let us not leave it at that.

Technically, the issue is that $\sqrt{g}\,R^{\mu\nu}(g) = \lambda g^{\mu\nu}$ needs to be fulfilled. However, "the only way to get $\sqrt{g}$ from flat space is to put in by hand a source $\eta_{\mu\nu}$ of the initial linear equations" (Deser, 1987, p. L103), which "violates the assumed fall-off at infinity of the h field" (Deser, 1987, p. L103). To be able to construct a cosmological term in the spin-2 framework, we need to start out from a more general – curved – background space, namely from an *Einstein space*[24] (with a Lorentzian signature). Deser (1987) and Deser and Henneaux (2007) show that the derivation presented for a flat (i.e., $R^{\mu}_{\nu\sigma\rho} = 0$ everywhere) background η can indeed be generalised to the class of curved backgrounds $\tilde{g}$ with $R_{\mu\nu} \propto \tilde{g}_{\mu\nu}$. This permits the construction of the cosmological term (Deser, 1987). Note, however, that this again might conflict with arguments against underdetermination that try to bring out the flat Minkowski background as preferred, as I myself argue below.

4.1.4.2 *Topological and Causal Restrictions*

Second, given these assumptions about the background manifolds from which we start, one might worry whether certain topological or causal restrictions on the solution space can enter: although, formally, the full action of general relativity is obtained, the solution space – i.e., the space of recoverable g fields – might differ from full general relativity due to such inherited restrictions. The situation is further complicated by the fact that the spacetime one starts with is not the spacetime one ends up with. It is, *prima facie*, not clear whether such restrictions at the start are inherited by the result.

Some issues are rather certain, though. For instance, moderate constraints for h may enter when one demands the retention of the correct signature of the effective metric g, or the correct relation of the two respective null cones (such that the effective g-causality does not violate the background η-causality; see Penrose, 1980; Pitts & Schieve, 2001a; 2001b). The latter arguably results in a restriction to *globally hyperbolic*[25] solutions – a deviation from standard general relativity that could in principle be tested empirically (Pitts & Schieve, 2001b).[26] It is important to note that it is standardly believed that "all physically realistic spacetimes must be globally hyperbolic" (Wald, 1984, p. 202).

A standard example for what is excluded by the restriction to globally hyperbolic solutions are spacetimes with closed timelike curves (Wald,

1984, p. 205). Accordingly, Gödel's (1949b) solution for a stationary, homogeneous universe with negative cosmological constant that is filled with rotating dust (Dautcourt & Abdel-Megied, 2006, p. 1269; Thorne, 1993, pp. 296–297) cannot be accommodated by spin-2 theory – the dust's rotation "tilts" the light cones such that closed timelike curves appear, which "pass through every event", since this "spacetime is stationary and homogeneous" (Thorne, 1993, p. 297). Gödel's solution is taken to be unphysical, because our universe has a small positive cosmological constant and (approximately) zero rotation. Similarly, Kerr and Kerr-Newman solutions (for, roughly, uncharged and charged rotating black holes, respectively) cannot be accommodated in full, since closed timelike curves may appear in certain regions. It is usually argued that in physically realistic Kerr solutions, instabilities prevent the occurrence of closed time curves (Thorne, 1993, p. 304). As for Kerr-Newman solutions, again, the very features that make these solutions problematic for the spin-2 approach lead to the conclusion that they are unphysical anyway. All of these issues arguably require further investigation, but for now I shall assume that one need not worry too much about such restrictions of the solution space.

Note, however, that one might wish to keep such supposedly unphysical solutions for other reasons. After all, the existence of such solutions is frequently used in metaphysics to determine the *modal* status of various structures: building on the famous apparatus of possible worlds, some structure is *necessary*, if it features in all solutions (or models) of a theory, and it is *contingent*, if it features only in a proper subset of them.[27] We may add that a structure is *forbidden*, if it is absent in all solutions.

For instance, Gödel's solution can be taken to suggest that time *order* (unlike time *direction*) is not a necessary property of time (e.g., Bartels, 2002; Friebe, 2012; Yourgrau, 1999). Instead, whether a universe has a universal time order depends on certain empirical facts about the distribution of matter. Notably, Gödel (1949a) himself took his solution to indicate that the universe is fundamentally timeless. Such arguments rest on the assertion that solutions like Gödel's are logically perfectly consistent. It is just that these solutions do not apply to our actual universe. *If*, however, the conditions for such solutions *were fulfilled* – for example, if cosmic dust were rotating fast enough relative to our local inertial frame – then closed timelike curves *would be* physical. In this sense, the metaphysics of time, for example, depends not only on the realistic models of our actual universe, but on all models, including the most bizarre.

There might still be reasons to exclude such solutions as purely mathematical. In fact, attributing physical significance to a solution is often tied to imposing additional constraint conditions, like energy conditions (see Curiel, 2017); notice, though, that energy conditions specifically do not exclude closed timelike curves as unphysical.

4.1.4.3 *Analytic Solutions*

Third, and more as a clarifying remark, one might worry that deriving the Einstein field equations perturbatively restricts the solution space to analytic solutions (i.e., solutions expressible by a convergent power series). It may be that some approaches, like Gupta's (1954), are affected by this. However, most classical derivations are not perturbative in the sense of adding orders of a small perturbation; in particular, Deser's closed form derivation is not, and neither is Kraichnan's (1955). Accordingly, no restrictions should apply.

For derivations in the quantum context the situation will generally be different (see Section 5.2). This is not surprising, though, since we certainly do not expect to retain all features of general relativity in a quantum framework. After all, both frameworks – quantum theory and relativity theory – impose restrictions. In particular, not all solutions that are compatible with the principles of quantum theory will conform with relativistic principles and *vice versa*.

4.2 The Dynamical Approach to Classical Spin-2 Gravity

Let us briefly retrace our steps. We have seen that it is possible to arrive at general relativity with its dynamical curved metric field g from a classical field theory of a massless spin-2 field h on a fixed flat Minkowski background η. At heart, i.e., with respect to the fundamental assumptions, general relativity still is a special-relativistic theory. On the spin-2 view, it seems that even the spacetime fundamentalist has to accept g as, in fact, ontologically *dependent*, i.e., non-fundamental: according to a fundamentalist reading of spin-2 theory, g ontologically depends on two ontologically independent entities, the fixed Minkowski field η and the dynamical spin-2 field h. This amounts to a fundamentalist view of the GA-SR type (see Section 3.5.2).

4.2.1 *The Dynamical Approach Resurrected*

However, an ontologically more radical and conceptually more coherent reading of the non-fundamentality of g is almost self-evident. Recall that fixed-background theories (like special relativity) allow for a complete ontological reduction of the fixed background to symmetry properties of the matter field dynamics: the dynamical approach to special relativity regards the Minkowski metric η as non-fundamental, namely as ontologically dependent on the Lorentzian symmetry properties of the matter field dynamics. Since specifying the Lorentzian symmetry properties of the matter field dynamics completely determines the Minkowski metric, insisting on adding the Minkowski metric to the fundamental ontology would amount to introducing an ontological redundancy.[28]

This can be used to reinterpret the non-fundamentality of g, and, thereby, provide a *fully dynamical* picture of spin-2 theory (and hence general relativity): the fixed Minkowski background η, which features in the spin-2 approach, is derivative on the Lorentzian symmetry properties of the matter field dynamics *including the new field h* (whose dynamical symmetry properties are also Lorentzian). Therefore, g does *not* simply ontologically depend on η and *h* as the proponent of the geometrical view wants to read the identification g = η + *h*. Instead, the dynamical understanding of spin-2 theory suggests that the g field is, in fact, ontologically dependent on the totality of Lorentz-invariant matter field dynamics including *h*. More precisely, g is completely ontologically reduced to the dynamical field *h* and the Lorentzian dynamical symmetry properties of all matter fields including *h*. Hence, g is *not* part of the fundamental ontology.[29] The spacetime structure of our world is effectively general-relativistic *because of the h field and the Lorentzian dynamical symmetry properties of all the matter fields (including h)*. The Einstein field equations and the g field provide an effective higher-level description of the underlying special-relativistic matter field dynamics.

Both the chronogeometric and the gravitational significance[30] of g are accounted for by the ontological reduction: g inherits the dynamical properties of *h* and the other matter fields and thereby obtains its capacity to act as the effective spacetime metric. This strengthens the dynamical approach to general relativity and distinguishes it from the tenable geometrical approach QGA. Recall that Read (2020) argues that DA-GR is indistinguishable from QGA (see Section 3.5.3). In particular, the second miracle is explained (see Section 4.2.2). We are left with a position that only commits to the fundamentality of matter fields, i.e., arguably some version of relationalism (see Pooley, 2013).[31]

Note that there remains a sense in which some form of spacetime fundamentalism is feasible: the *h* field might be considered as spatiotemporal. After all, it couples universally and builds g. Since *h* is also a plenum, it might even be taken to represent spacetime itself. The spacetime fundamentalist may therefore change horses and agree that g is not fundamental, but point out that *h* is. However, *h* can be zero – Maudlin (1993) explicitly excludes such fields as an option for the spacetime fundamentalist. The relationalist may defend her position by, for example, stressing that identifying spacetime with *h* is the least plausible option for the spacetime fundamentalist (g and η are more natural choices) – *h* can be zero, *h* does not have a Lorentzian signature, *h* carries energy, *h* interacts with itself, and so on; this quickly becomes a purely semantic dispute, though. What is more, the most plausible alternative to relationalism here is, in fact, spacetime emergentism (see Section 2.2.4 and Martens, 2019). Put crudely, the spacetime emergentist's perspective on the spin-2 approach is that spacetime is shown to "emerge" from the underlying fundamental posits, and in particular from *h*. Unlike a modified form of spacetime fundamentalism, spacetime emergentism does justice to the fact that g is without doubt the effectively observed spacetime. I shall come back to this option in the context of quantum gravity in Chapter 5, where spacetime emergentism is arguably even more preferable to brute spacetime fundamentalism.

But there is another sense in which spacetime fundamentalism is not ruled out. The dynamically interpreted spin-2 theory suggests an understanding of general-relativistic spacetime as a *partly* non-fundamental structure: as presented, specifically *metrical* properties of spacetime are non-fundamental. What remains untouched, however, are – again – the *topological* properties of spacetime, i.e., the manifold. The problem of pregeometry reappears just as pressing as for the standard Brownian dynamical approach; and the possible solutions are the same as the ones presented in Section 3.6.

I have argued, and will continue below, that the dynamical view is strengthened by the spin-2 approach to general relativity. This support is two-way. Not only does the dynamical view benefit from the spin-2 approach to general relativity, but the spin-2 approach also benefits from the dynamical view. Although I generally take the view that the foundational and philosophical importance of the spin-2 approach is not tied to the dynamical interpretation, it is the dynamical view that provides the ontologically most parsimonious and conceptually most coherent interpretation of the spin-2 approach to general relativity. In particular, the dynamical view helps to resolve potential reservations against the spin-2 approach to general relativity.

For instance, Pitts (2016) raises doubts about claims to the effect that the ontology of the spin-2 approach is obvious, and that the spin-2 derivation of Einstein's field equations shows that general relativity simply becomes another special-relativistic field theory. Pitts (2016, p. 64) argues that it is not clear whether the "unobservability" of the flat metric (Kraichnan, 1956, p. 484; Thirring, 1961, p. 96), from which one starts in the spin-2 approach, accounts for its "not being real" (which Pitts takes to be tacitly assumed), and, hence, for its "ceasing to exist" in favour of the emergence of the effective curved metric field g:

> Eventually it is concluded that the flat background metric is "unobservable", which usually is supposed to mean or at least to imply that it doesn't really exist, perhaps (Thirring, 1961). But how does coalescence of the flat geometry and the gravitational potential make the flat geometry cease to exist?
>
> (Pitts, 2016, p. 64)

Pitts' point draws attention to the following: as mentioned at the beginning, spin-2 theory is standardly (i.e., in a geometrical interpretation) committed to *two* fundamental metric fields, g *and* η, with only g being physically significant. There is in principle no theoretical problem with having more than one metric field, but it is arguably counter-intuitive (although Pitts is affirmative of this view). Accordingly, it casts doubt on why one should prefer the spin-2 view to the standard interpretation of general relativity, which only postulates a single metric field. Without further argumentation, one cannot simply argue that the fact that η is unobservable *in principle*, justifies applying Occam's razor, since η seems to be an important part of the reductive explanation of g.

However, this potential problem for the spin-2 approach virtually disappears, when we use Brown and Pooley's understanding of the dynamical approach to special relativity: on their view, the flat background is a non-entity *from the start* – it does not "cease to exist". Moreover, its being a non-entity is justified not by its being unobservable, but by its being *redundant*, which, to Brown and Pooley, makes the Minkowski metric not only non-fundamental, but also not real. The Minkowski metric is a mere codification of the Lorentzian dynamical symmetry properties of the matter fields.

This argument is not available to the proponent of a geometrical reading of the spin-2 approach to general relativity. Thus, the dynamical approach perspective increases the coherence of the spin-2 approach. A proponent of this understanding of the dynamical approach, who equates non-fundamentality with non-existence, will probably also dismiss g as a non-entity – only the matter fields are real.

But what if one understands the dynamical approach to be primarily concerned with questions of fundamentality rather than existence, as I myself have repeatedly defended in this work? Then Brown and Pooley's verdict on the non-existence of the Minkowski metric is not available. Consequently, it seems that the (then merely non-fundamental) Minkowski metric *is* part of the overall ontology, just as the g field is. Both metric fields seem to exist, namely non-fundamentally. In fact, it even seems that the geometrical approach is suddenly advantageous, since here η and g *do not share the same ontological status* (in contrast to the dynamical view): both metrics are in the ontology, but η is fundamental while g is derivative. As a result, the proponent of the geometrical view can seek to argue that their interpretation is more coherent, since it ontologically reflects that η is part of what explains (or builds) g; and, that their interpretation also better reflects that only g is physically significant: it is rather common that we do not have direct empirical access to the fundamental constituents, but only to the derivative entities they build (e.g., tables and their fundamental constituents). Note, however, that η's being inaccessible *in principle* is arguably not what we are typically prepared to accept (even quarks do show indirectly). Moreover, the proponent of the dynamical approach can counter that derivative objects can of course explain and build other derivative objects (e.g., atoms build and explain tables).

Still, it is fair to say that Pitts' remark becomes pressing once Brown and Pooley's non-entity argument is unavailable. So, one might ask, why prefer the spin-2 view to the standard view of general relativity, which only postulates a single metric field g?

Here is why. The above issue is, in fact, only an apparent problem for my view of the dynamical approach. It is not Brown and Pooley's understanding of the dynamical approach to special relativity (which centres on the issue of existence) but my understanding of the dynamical approach (which centres on the issue of fundamentality) that makes clear *why* η is not real in the spin-2 approach to general relativity independently of general metaphysical issues regarding whether or not derivative objects are real. Consider the following argument. If God had *explicitly* created all Lorentz-invariant matter fields *except* the spin-2 field *h*, then she would have *implicitly* created the Minkowski metric (as a derivative object), and the world would be described by special relativity. However, if God had explicitly created all Lorentz-invariant matter fields *including* the spin-2 field *h*, then she would have *implicitly* created the effective g field (as a derivative object), and the world would be described by general relativity – period. At no point is the Minkowski metric of concern in the latter case, hence, at no point does the Minkowski metric exist in the latter case. The assumption to the contrary results only from overemphasising the ontological import of the concrete mathematical derivation which features $g = \eta + h$. Consider the following analogy. If I mix flour, butter, and sugar, and bake it, I obtain a traditional British pound cake. If I start with the same ingredients plus lemon juice, I obtain something else, namely a lemon cake. I do not end up with both cakes, but one – only the lemon cake is real (albeit non-fundamental).

Therefore, the derivative metric g is the sole metrical structure in the spin-2 approach to general relativity. It is on my understanding of the dynamical approach, which focuses on the issue of fundamentality rather than existence, that the dynamical approach explains *why* there is only one metric, i.e., *why* η is a non-entity in the dynamically interpreted spin-2 approach to general relativity, independently of a general metaphysical stance on the existence of derivative objects. Thereby, the dynamical approach increases the coherence of the spin-2 approach to general relativity – and *vice versa*.[32]

All this is not sufficiently appreciated by Brown (2005), who only briefly mentions the possibility of a spin-2 derivation, without properly acknowledging its strength or its usefulness for his programme. Brown notes that "[i]t is widely known that there is a route to GR [general relativity] based on global Minkowski spacetime with spin-2 gravitons" (Brown, 2005, p. 172), but (1) takes it that this involves "requir[ing] that $h_{\mu\nu}$ couple to its own stress-energy tensor, as well as to the matter stress-energy tensor" (Brown, 2005, p. 172). This downplays the most significant aspect of the derivation, namely that the coupling properties of h, which are key for reproducing the general-relativistic g field as the effective spacetime metric, result from merely demanding mathematical consistency. *For consistency reasons*, an interacting spin-2 theory is forced to reproduce general relativity. Furthermore, Brown (2) does not make it explicit that this can be read as providing a fixed-field formulation of general relativity that allows for similar ontological conclusions as in the case of special relativity; and, (3) does not discuss potential issues that arise when taking this "route to general relativity" seriously (for example, regarding its equivalence to general relativity). Brown does point out, though, that "[t]he flat background metric in this approach becomes bereft of any direct operational meaning because the SEP [strong equivalence principle] holds for the local inertial frames defined relative to $g_{\mu\nu}$ and not to $\eta_{\mu\nu}$" (Brown, 2005, p. 172). But, in line with my second comment, this does not give sufficient credit to Brown's own core idea regarding the dynamical approach to special-relativistic scenarios, nor to my variation of it in the spin-2 context: that the flat background is a non-entity.

4.2.2   A Miracle Disappears

There is more to say about my claim that the dynamical approach to general relativity is strengthened by the spin-2 view. Essentially, the spin-2 view increases the explanatory strength of the dynamical approach to general relativity. Recall that Read et al. (2018) and Read (2020) argue that two unexplained "miracles" appear in general relativity (see Section 3.5.4). The first miracle concerns the fact that the dynamical symmetry properties of all matter fields are Lorentzian; the second miracle concerns the fact that the dynamical symmetry properties of the matter fields locally coincide with the symmetry properties of the metric field g. Notably, they argue that the miracles arise *independently* of endorsing a particular view in the debate on the dynamical approach, as a fact about general relativity itself. The question is whether this is correct. Is the failure to explain the two miracles really a defect of general relativity itself, which can only be resolved in a successor theory – as, for example, Read (2019) argues?[33]

The results of this chapter cast doubt on that. Classical spin-2 theory – which is not a successor theory, but can be viewed as equivalent to general relativity (up to the qualifications discussed above) – already provides what is required to explain the second miracle: a reconceptualisation of general relativity that, in a dynamical reading, allows for a complete ontological reduction of general relativity's metrical structure to matter field structure. As a result, the second miracle is *not* generally miraculous in general relativity, but only within a (tenable) geometrical interpretation, from which the dynamical approach is thereby clearly distinguished – *pace* Read (2020).

The argument is two-step. First, the fundamental ingredients of the dynamically interpreted spin-2 theory are the matter fields (including *h*) and their Lorentz-invariant dynamics. The effective metric field g is derivative on these fundamentals: explicitly creating the fundamentals implicitly creates g. The g field inherits the properties of the dynamics it ontologically depends on: the dynamical symmetry properties of all the matter fields (including *h*) and the universal coupling property of *h*. As a result, on a dynamical reading, the symmetry properties of the derivative object g "coincide" *trivially* (or, as Read, 2020, p. 183 would put it: "automatically") with the dynamical symmetry properties of the matter fields (including *h*). This is because the latter completely *determine* the former. Accordingly, also g's chronogeometricity follows automatically: the second miracle is explained in spin-2 theory.

The specific type of dynamical symmetry properties of the matter fields is, of course, assumed; it happens to be an empirical fact that the dynamical symmetry properties of all matter fields are Lorentzian. So, a dynamical reading of spin-2 theory takes MR1 as an unexplained assumption and explains MR2 in terms of a complete ontological reduction – similar to the case of special relativity.

In a second step, this result from spin-2 theory is then attributed to general relativity by appeal to the equivalence claim that the full non-linear spin-2 theory is, in fact, general relativity. Again, chronogeometricity is then *better explained* than in the standard Brownian dynamical approach to general relativity, where MR2 is miraculous.

It needs to be flagged, that, in light of Section 4.1.4, spin-2 theory is not strictly equivalent to standard general relativity in the following sense: spin-2 theory cannot accommodate *all* g fields that are in the solution space of general relativity. There are solutions of general relativity which cannot be reconstructed in the spin-2 approach. Accordingly, the "spin-2 argument" is not applicable to explain why the dynamical symmetry properties of *such* g fields coincide with the dynamical symmetry properties of the matter fields – for such g fields, MR2 can*not* be explained. In other words, there may be possible (or, conceivable) worlds where the results of spin-2 theory do not translate to general relativity, so that the second miracle remains unexplained.

Nevertheless, MR2 can be explained for all those solutions of general relativity that are also solutions of spin-2 theory. In particular, the result presumably holds for all empirically relevant solutions of general relativity. Hence, I do not take this shortcoming of spin-2 theory to vindicate the conjecture by Read et al. (2018) and Read (2020) that MR2 can only receive an explanation from a successor theory, i.e., from "outside" general relativity: spin-2 theory and general relativity are not so much "different" theories in the sense that their central equations are different, but these equations have different solution spaces. In other words, the argument from spin-2 theory is an argument *within* the limitations of general relativity, it just does not apply to the full solution space of general relativity. Thus, whether MR2 can be explained or not is rendered a fact about the particular solution and hence, in a sense, the particular *world*, not the theory.

At the heart of the spin-2 argument lies the fact that, contrary to previous verdicts on the dynamical approach to general relativity, it is actually possible to separate the dependent and independent structures, that is, the dynamical and fixed parts of g in order to obtain a fixed-field formulation of general relativity. Generally, fixed-background theories permit a complete ontological reduction of the fixed field, such that MR2 does not appear miraculously. In this sense, the success of the spin-2 argument can be anticipated (which does not render it uninteresting, of course). In light of the spin-2 argument, the second miracle only seems unexplained in general relativity due to an obscuring representation of the theory – which, ironically, is the standard representation.

If the two formulations are (almost) equivalent, i.e., do not represent (substantially) different theories, all conceptual aspects should translate. In particular, there should be a *fact* about whether general relativity is fully dynamical or has a background. The solution to this is to appreciate that the fixed part is redundant to the dynamical symmetry properties of the matter fields. It does not feature in the theory as a real background field, but is a non-entity, as discussed above. Thus, the dynamical approach to spin-2 theory has additional import for the issue of *background dependence* usually raised for such "fixed-background" theories. In brief, the standard lore is that one of the most important features of general relativity is that the metric field g is dynamical, such that the theory is background-independent.[34] In contrast, the spin-2 approach seems to have such a fixed background, namely the Minkowski metric η. Hence, the theory may seem inferior to general relativity on these grounds. Now, there are several caveats to this already at the level of a geometrical interpretation. First and foremost, one needs to properly define the notion of background independence; depending on the notion used, general relativity may or may not be background-independent (Read, 2016). It also depends on whether spacetime is empty or not (Belot, 2011). However, the proponent of the dynamical approach to general relativity can additionally argue that this "fixed background" simply expresses an empirical fact: Lorentz symmetry. That it is "fixed" is just the first miracle, which holds in standard "background-independent" general relativity, too. Hence, in a dynamical interpretation, either both theories or neither of them are background-independent. Notably, it is irrelevant whether MR1 is explained or not.

There is another important aspect to the issue of background independence that deserves further scrutiny: one could choose another (non-Minkowskian) background. In other words, one could choose a different "split" $g = \tilde{g} + h$. Essentially, the first miracle amounts to why $\tilde{g} = \eta$ is the most natural choice, or *best-system choice*, for the initial background in spin-2 theory – or, more precisely, for what the background codifies in a dynamical reading: the dynamical symmetry properties of h and the other matter fields. On the dynamical view, the second miracle is then automatically explained. But on a geometrical reading, the best-system choice brings in both miracles: it is only for this specific choice that the symmetries of $\tilde{g}$ and h coincide. This observation is a variant of what Read (2020) emphasises with respect to the tenable geometrical approach to special relativity (see below).

To conclude, general relativity understood in light of a dynamical reading of the spin-2 approach demonstrates that local dynamical symmetry properties of matter fields do not miraculously coincide with the symmetry properties of the g field. They do so because the matter fields ontologically "build" the g field – the g field ontologically depends on the matter fields and hence obtains its symmetry properties from them. In a sense, MR2 is reduced to MR1: the coincidence of symmetry properties is derived from assuming that all matter field dynamics are Lorentz-invariant.

With respect to the geometrical–dynamical debate, this suggests that the dynamical approach to (a spin-2 understanding of) general relativity is different from and preferable to the (qualified) geometrical approach to (a spin-2 understanding of) general relativity, for essentially the same reasons that Read (2020) puts forward in the case of special relativity: the dynamical approach is preferable because (a) it explains more – the dynamical approach only has to accept MR1 as unexplained, while QGA has to accept both miracles as unexplained[35]; and (b) the dynamical approach has a "simpler" ontology than QGA.[36] Hence, it is the spin-2 approach to general relativity that enables the dynamical approach to fully deliver on its explanatory promises and that clearly distinguishes it from the rival geometrical position.

4.3 Fundamentality of Spacetime Revisited

I have argued that the following lessons can be drawn from the spin-2 approach to general relativity: (1) general relativity is reduced to or derivable from classical spin-2 theory; (2) this allows for a full resurrection of the dynamical understanding of spacetime in general relativity; and (3) in light of the foundational insights from classical spin-2 theory, the so-called second miracle of general relativity is not miraculous, but explained as a consequence of the first "miracle" which is introduced to the theory as an unexplained brute empirical fact.

In contrast to the standard Brownian dynamical approach, these results pave the way to an understanding of general-relativistic spacetime, or rather *key aspects* of general-relativistic spacetime (namely, metrical aspects), as non-fundamental. Thereby, the dynamical interpretation of the spin-2 approach foreshadows current developments in the context of (speculative) theories of quantum gravity (see Chapter 5) and demonstrates that such research programmes are, in fact, more continuous to established theories, even with respect to ontology, than usually acknowledged. In particular, the widespread assertion that spacetime will ultimately turn out to be non-fundamental no longer draws on speculations about new physics alone, but receives additional support from a re-interpretation of *well-known* physics.[37] This considerably strengthens spacetime non-fundamentalism.

With only the metrical aspects of spacetime being reduced to matter field dynamics, dynamical spin-2 theory does not yet imply a full-blown constructivist understanding of general-relativistic spacetime. The problem of pregeometry also applies to the spin-2 approach. However, solution strategies are available, as I have argued in Section 3.6.

That said, there are three issues that might push back on the findings of this chapter, and especially their ontologically conclusiveness. First, we have seen that assuming Lorentz invariance of matter field dynamics (i.e., MR1) is at the heart of the argument I presented. That the matter field dynamics are Lorentz-invariant is a posit that remains unexplained and needs to be accepted as a brute empirical fact. The laws of physics just happen to exhibit this symmetry. But do they? One might wonder whether taking Lorentz invariance as the empirical basis for the foundational issues tackled here is sufficiently justified. This seems especially important, because non-Lorentzian matter fields, if existing, would still enter the Einstein field equations via the energy-momentum tensor – which is precisely what Read et al. (2018) and Read (2020) use in their problem case rationale against the (untenable) geometrical view. Thus, one might ask, what if Lorentz invariance is violated or turns out to hold only effectively, i.e., is violated in the fundamental theory? In fact, some – but not all – theories of quantum gravity require Lorentz invariance to be violated (see Chapter 5). Regarding empirical data, however, no violations of Lorentz invariance have yet been found (Mattingly, 2005). So any violation of Lorentz invariance is expected to be a high-energy effect. Furthermore, as Carlip (2014) points out, with reference to Collins et al. (2004), "even violations at very high energies can feed back into quantum field theory through loop effects and lead to drastic consequences at low energies" (Carlip, 2014, p. 203). Thus, the empirical results for Lorentz invariance are further strengthened. In addition, small violations of Lorentz invariance are considered to "lead to problems with black hole thermodynamics" (Carlip, 2014, p. 203). To conclude, Lorentz invariance is an empirically well-established and robust basis for foundational arguments in physics.

Second, why should one commit to what some might consider an "ugly way" of thinking about general relativity, especially when, in the end, it gives rise to g?[38] I would generally opt for not getting distracted by notoriously vague concepts like "beauty" (cf. Hossenfelder, 2018). Instead, we should appreciate that a proper assessment is about issues of explanation and fundamentality. There is no doubt that the objects of statistical mechanics are more fundamental than the objects of thermodynamics, although thermodynamics may often provide the more compact or, if one wishes, "beautiful" description. More concretely, one can counter that, first of all, the spin-2 approach obviously needs to give rise to g (at least approximately) – otherwise it would fail empirically. Then, as pointed out before, the dynamically interpreted spin-2 approach is preferable, because it is the explanatorily most coherent interpretation of general relativity – it assumes less unexplained facts and better explains how g obtains its chronogeometricity; again, the dynamical approach and spin-2 theory support each other mutually: combining the two to what I dubbed the "dynamical spin-2 approach (to general relativity)" is explanatorily more coherent than the three alternative combinations available: (1) combining the (Brownian) dynamical approach and standard general relativity to the "standard (Brownian) dynamical approach", (2) the geometrical approach and standard general relativity to the "standard geometrical approach", and (3) the geometrical approach and spin-2 theory to the "geometrical spin-2 approach".

However, with respect to the fundamental ontology, the situation looks less clear-cut: while the dynamical spin-2 approach is ontologically more parsimonious than the geometrical spin-2 approach, it is merely equally ontologically parsimonious (quantitatively) as either interpretation of standard general relativity that is committed to g instead of h. With respect to the fundamental ontology, the dynamical spin-2 approach seems to be on a par with the standard formulation of general relativity (in either interpretation). To counter this, one may seek to argue that the dynamical spin-2 approach commits to a *less specific* or *more generic* fundamental ontology: committing to "just another" matter field, even though one with the specific properties of being a spin-2 field with globally Lorentz-invariant dynamics, and then *deriving* its physical implications is arguably more generic than committing right away to these physical implications as properties of g. A theory which shows that a rather specific entity ontologically depends on more generic structures could then be argued to be preferable for having the *qualitatively* more parsimonious fundamental ontology. This move especially tells against the draw with the standard geometrical approach, where g is arguably a rather specific entity. The standard dynamical approach seems less affected by such an argument, though, since it takes g to be "just another" field – just like the dynamical spin-2 approach does with respect to *h*. Note, however, that for the latter this reasoning is *manifest*, whereas for the former it is merely *rhetorical*: after all, even the proponent of a geometrical view would *not* count *h* as spatiotemporal in spin-2 theory – they would clearly argue that it is η which represents spacetime, whereas *h* is just some field defined on η. The same is, of course, not true of g in the standard dynamical approach. In this sense, the fundamental ontology of the dynamical spin-2 approach can be considered quantitatively on a par with, but qualitatively more parsimonious than both interpretations of standard general relativity, and hence preferable.

However, and here is the third objection, why should we take the spin-2 approach as having this import for questions of fundamentality in the first place? Why is it not just a change of representation? Or similarly, how should we decide which of the pictures is true of the world?

In fact, Feynman et al. (1995, pp. 112–113) implicitly (see also Pitts, 2017, p. 265) and Sexl and Urbantke (2002) explicitly conclude that the spin-2 results confirm an analogue of the famous charge of conventionalism raised by Poincaré (1905, Chapter IV–V) against non-Euclidean geometries – notably, prior to general relativity. Note, however, that Poincaré did not argue for an underdetermination, i.e., that we cannot settle for what is true of the world, but that such findings indicate that there is no such thing as spacetime geometry at all. Poincaré takes our freedom to choose a representation conventionally as proof that it is really nothing but a convention without ontological impact. While this position historically advocated by Poincaré may be avoidable, the underdetermination version is not so easily.

Indeed, there are reasons to doubt that classical spin-2 theory on its own is conclusive with respect to re-evaluating the fundamental ontology. Here is why. To make an ontologically clear case for g being non-fundamental, i.e., for g being ontologically dependent, we would need to argue that g is determined by η and *h* – *but not vice versa*. One can raise severe doubts about whether this is possible in classical spin-2 theory. This is due to the symmetry of $g = \eta + h$: g may be viewed as ontologically depending on η and *h*, but, likewise, *h* may be viewed as ontologically depending on g and η. Just showing that we can "split" g into these "parts" does not do the job. In a sense, "parts" and "whole" determine each other mutually.[39] Classical spin-2 theory has arguably established that there is *some kind* of ontological dependence relation between g, η, and *h*, but without further interpretation (like the one given above, for example) it does not establish the direction of this dependence relation or whether there is a preferable direction at all.[40] In fact, why not interpret $g = \eta + h$ as the symmetric identity relation it mathematically is? Of course, starting from a field theory for h and ending up with g and general relativity seems to suggest otherwise, but, so the reasoning goes, we may equally well argue that it is g which is fundamental: mathematically, we can just run the whole derivation backwards such that h is obtained from explicitly "splitting" g (see Appendix A.1); and, empirically, g is the relevant object anyway. So why should g be derivative and h fundamental?

This problem is particularly pressing in the geometrical view: considering both h and η as ontologically independent "parts" of an ontologically dependent "whole" g has no merit at all compared to viewing g as fundamental, with regard to explanatory and ontological considerations; in fact, a fundamental g is *more* parsimonious.[41]

However, this changes when adopting the dynamical view. On this view the dependence relation $g = \eta + h$ can*not* be interpreted as a simple part–whole relation anymore because η and h are *not* ontologically independent entities[42]: η ontologically depends on symmetry properties of h, it merely reflects that h and the other matter fields are Lorentzian.[43] In fact, we can shortcut this argument by pointing out that η is a non-entity (see Section 4.2.1). As a result, the dynamical view suggests to not interpret the mathematical derivation literally.

This and the fact that the dynamical interpretation of spin-2 theory (with h being fundamental) is explanatorily more coherent (fewer miracles, better explanation of g's chronogeometricity) and ontologically more parsimonious (fewer fundamental entities) *does* suggest that it is h which should be considered fundamental, and that it is g which should be considered derivative. Hence, it is the dynamical view that enables the spin-2 approach to most convincingly establish that h is fundamental and g is derivative, i.e., that metrical aspects of spacetime are non-fundamental.

Notice that although this argumentation differs from the one regarding special relativity, it again hinges on the interpretation of η. So why should the spacetime fundamentalist suddenly accept such reasoning? Arguably, the spacetime constructivist (or the proponent of the dynamical spin-2 view, for that matter) could use additional support in order to break what at least the fundamentalist wants to see as a tie and establish itself as the clearly better position. Here, the research programme of quantum gravity and, more specifically, the existence of a *quantum* spin-2 approach might help. For now, it is key to note that taking metrical aspects of general-relativistic spacetime as non-fundamental is at least as *compatible* with classical general relativity as a fundamentalist interpretation.

Notes

1 This chapter is based on work previously published in Salimkhani (2020). The text has been substantially revised and extended and is reprinted with permission from Elsevier.

2 For example, Markus Fierz, and Wolfgang Pauli (1939), Nathan Rosen (1940a; 1940b), Achille Papapetrou (1948), Suraj Gupta (1952a; 1952b; 1954; 1957), Robert Kraichnan (1955; 1956), Walter Thirring (1959; 1961), Ogievetsky and Polubarinov (1965), Walter Wyss (1965), Fang and Frønsdal (1979), Feynman et al. (1995), Steven Weinberg (1964a; 1964b; 1972), Stanley Deser (1970; 2010), Robert Wald (1986), and Pitts and Schieve (2001c; 2007).

3 As Lorentz indices μ and ν run from 0 to 3 in integer steps, $X_{\mu\nu}$ is a 4×4 matrix with 16 components.

4 In representation theory, a *representation* is a mapping $e \to D(e)$ of the group elements e to a set of matrices $D(e)$ such that the group operation is preserved, i.e., if $ab = c$, then $D(a)D(b) = D(c)$. A representation is *reducible*, if there exists a non-singular matrix M, independent of the group elements, such that it can be put in block-diagonal form, i.e.,

$$M D(a) M^{-1} = \begin{pmatrix} D_1(a) & & 0 \\ & D_2(a) & \\ 0 & & \ddots \end{pmatrix}$$

which can be written as the decomposition $D(a) = D_1(a) \oplus D_2(a) \oplus \ldots$. If this is not possible, $D(a)$ is an *irreducible* representation. For more details see, for example, Cheng and Li (1982) and Tung (2003).

5 Kupersztych (1988) may be understood as an illustration of this. Against the standard believe that, for example, "in classical electrodynamics, the spin of a particle seems to be an extraneous concept", which "seems only to be justified by the existence of the relativistic quantum theory", he shows that the notion of electron spin automatically appears when "using only the laws of classical electrodynamics (Maxwell equations, Lorentz force law) and invariance properties (gauge invariance, relativistic invariance) in the problem of an electron interacting with the field of an electromagnetic plane wave" (Kupersztych, 1988, p. 448).

6 Similarly, any additional internal symmetry will lead to additional numbers of classification, for example, electric charge q from $U(1)$ gauge symmetry.

7 For example, the numbers m and s – for mass and spin, respectively – are sufficient to classify all possible (continuous, unitary, and non-negative energy) irreducible representations of the Poincaré group. It is important, though, to note that the (homogeneous) Lorentz and the Poincaré group do *not* share the same Casimir invariants, since the relation of the two groups is non-trivial – in particular, the Poincaré group is not merely a direct product of the Lorentz group and the group of translations. For the Lorentz group, with generator $M_{\mu\nu} = L_{\mu\nu} + S_{\mu\nu}$, the Casimir invariants are $1/2\ M^{\mu\nu}M_{\mu\nu}$ and $1/4\ \epsilon_{\mu\nu\kappa\lambda}M^{\mu\nu}M^{\kappa\lambda}$ (Noz & Kim, 1986, p. 78). On the other hand, the Casimir invariants of the Poincaré group are $P^2 = p^{\mu}p_{\mu}$, with the four-momentum p_{μ}, and $W^2 = W^{\mu}W_{\mu}$, with the Pauli-Lubański pseudovector W_{μ} describing the total spin of a massive particle (Noz & Kim, 1986, p. 61). The irreducible representations of the Poincaré group are then labelled by the eigenvalues of the Casimir invariants which are associated with the mass of the particle by $m^2 = P^2$ and the spin of the particle by $W^2 = -m^2 s(s+1)$, respectively. Analogously, massless particles can be classified using helicity instead of spin.

8 Arguably, also the additional quantum-related restriction that extends the relevant symmetry group from the homogeneous Lorentz group (in the classical case) to the inhomogeneous Lorentz group (in the quantum case), and, specifically, the irreducible representations from finite-dimensional non-unitary ones (in the classical case) to infinite-dimensional unitary ones (in the quantum case), is responsible for spin being a fundamental property.

9 The spin-statistics theorem is a result of relativistic quantum theory. It states that the wave function $\Phi(x) = \varphi_1(x)\varphi_2(x)...\varphi_n(x)$ of a system of particles $\varphi_i(x)$ of the same species with integer spin (called bosons) is *symmetric* with respect to interchanging any two particles, i.e., $\Phi(x)$ *does not change* its sign; and that the wave function $\Psi(x) = \psi_1(x)\psi_2(x)...\psi_n(x)$ of a system of particles $\psi_i(x)$ of the same species with half-integer spin (called fermions) is *anti-symmetric* with respect to interchanging any two particles, i.e., $\Psi(x)$ *changes* its sign (for more details see Streater & Wightman, 2000). Note that according to quantum theory, any physical state is described by the square $|\psi|^2$ of some wave function or state vector $\psi \in \mathcal{H}$, where $\mathcal{H}$ is a Hilbert space. Therefore, any two state vectors ψ and φ that differ only by an overall phase $\vartheta \in R$, i.e., $\psi = e^{i\vartheta}$, describe the same physical state. In particular, any two state vectors that differ only by an overall minus sign describe the same physical state. So, interchanging two identical particles does not alter the physical state of the system.

10 One way to understand the spin-statistics theorem is in terms of the question whether two spacelike separated particles commute or anti-commute (see Massimi & Redhead, 2003; Streater & Wightman, 2000). This requires the light cone structure of special relativity. Sometimes this special-relativistic "locality" condition is referred to as "microcausality". Notably, Weinberg (1995) stresses that there is a sense in which the condition is not a locality or causality condition, but a condition that is needed to retain Lorentz invariance of the S-matrix (see also Massimi & Redhead, 2003).

11 Here, $\eta_{\mu\nu}$ is a fixed field in the sense of Pooley (2017, p. 115): it is identically the same in all kinematically possible models.

12 For reviews see, for example, Misner et al. (1973), Álvarez (1989), Preskill and Thorne (1995), Sexl and Urbantke (2002), Maggiore (2008), and Ortín (2015); original works are due, for example, to Gupta (1954), Kraichnan (1955), Deser (1970; 2010), Thirring (1959; 1961), Ogievetsky and Polubarinov (1965), Wyss (1965), Fang and Frønsdal (1979), Wald (1986), and Pitts and Schieve (2001c).

13 In a flat background setting, this standard field-theoretic assumption prevents the derivation of the cosmological constant term (Deser, 1987, p. L103).

14 See Preskill and Thorne (1995), and Fang and Frønsdal (1979) for an overview.

15 See Gupta (1952a; 1952b).

16 Furthermore, Deser (1970) does not need to introduce a particular gauge. See also Deser's (2010) reply to Padmanabhan (2008).

17 All attempts at deriving the Einstein field equations demonstrate that *only* by including all contributions (the 'full' theory) are the consistency conditions satisfied. Some (e.g., Deser, 1970; Fang & Frønsdal, 1979; Ogievetsky & Polubarinov, 1965; but not Gupta, 1954) even argue that this *uniquely* leads to general relativity – meaning that "Einstein's nonlinear theory of gravitation is the only consistent, Lorentz-invariant theory of an interacting, massless, spin-2 field in flat space" (Fang & Frønsdal, 1979, pp. 2267–2268). This stronger claim might even be considered the received view today (see Preskill & Thorne, 1995). However, such claims need to be evaluated carefully as, for example, Ortín (2015; 2017) argues. In particular, some additional presumptions will

typically be made (e.g., Ogievetsky & Polubarinov, 1965). In this work, I will not evaluate the uniqueness claims, but refer the reader to the original work due to Huggins (1962) and a later debate between Padmanabhan (2008), Butcher et al. (2009), Deser (2010), and Barceló et al. (2014).

18 See Linnemann et al. (2023) for a very recent critical take on whether general relativity can be derived from a self-interacting spin-2 approach. I take their results to be very interesting and thought-provoking but controversial. Amongst other things, it remains unclear to what extent this critique affects the merits of spin-2 theory for the foundations of general relativity and, especially, for metaphysics. But assume for now that their criticism of the physics holds – or, at least, that the subsequent philosophical points are sufficiently independent of it, as they state themselves. With regard to the reduction thesis defended in Salimkhani (2020) and this chapter, their key philosophical claim is presented as a dilemma: "either the view [of the metric field as a mere stand-in for a self-coupling spin-2 field in flat spacetime] is physically incomplete in so far as it requires recourse to GR [general relativity] after all, or it leads to an absurd multiplication of alternative viewpoints on GR" (Linnemann et al., 2023, p. 3). It strikes me that both horns of the dilemma are blunt. Generally, the first horn seems to rely on a strict understanding of what counts as permissible in such derivations. More details are needed regarding the question of what kind of "recourse to general relativity" is problematic. Not even being able to conceptualise an essential posit would arguably be severe, but facing the problem that there are many available energy-momentum tensors to choose from seems less so. In the latter case, the allegedly problematic recourse to general relativity (if there is any) may just be understood as a shortcut to empirical adequacy. I fail to see why using prior (empirical and theoretical) knowledge from general relativity (or any other part of physics) in the context of discovering spin-2 theory should break the spin-2 view – especially from a metaphysical perspective. Its derivation of general relativity may just be less impressive. To paraphrase Linnemann et al. (2023, p. 17): none of this undermines the heuristic use of general relativity. But to be clear: their genuinely foundational project is very helpful and important. The second horn has been addressed in Salimkhani (2020) – and is addressed in this chapter – via best systematisation and the explanatory and ontological import of the dynamical spin-2 perspective. Relatedly, the problem of underdetermination regarding fundamentality issues is addressed in detail in this book. Further criticism is presented by Darrigol (2007; 2014, Chapter 7) regarding, amongst others, concerns that the vanishing divergence of the total energy-momentum tensor is insufficient to determine the field equations. For the Gupta–Feynman approach Darrigol sees this extra assumption as one of what he dubs "Faradayan principles". In private communication Darrigol argued that Deser's approach uses a different potentially unjustified assumption. It may well be that such additional assumptions are invoked to constrain the mathematical machinery.

19 I thank Brian Pitts for helpful discussion.

20 Indeed, with respect to the ordinary (not the covariant) divergence (Weinberg, 1965b, p. 471).

21 To follow the rather brief exposition of the argument in the work at hand, as well as in the works cited, it is instructive to consider Weinberg's (1972, pp. 165–166) comments on how to separate linear from non-linear contributions of h in the Ricci tensor (see Appendix A.1).

22 In that case, one could still try to modify spin-2 theory; for example, by making the h field massive (e.g., Ogievetsky & Polubarinov, 1965; Pitts, 2017).

Note, however, that it has been argued that "spin 2 gravitons of finite mass, however small, are inconsistent with the observed light bending and present difficulties of principle as well" (Boulware & Deser, 1975, p. 194), so the massless case "must dominate the gravitational contributions to scattering amplitudes" (Boulware & Deser, 1975, p. 194).

23 See Smeenk (2013) for an introduction to the philosophy of cosmology.

24 An *Einstein space* is a generally curved (pseudo-)Riemannian differentiable manifold with the property that its Ricci tensor is proportional to its metric, i.e., $R_{\mu\nu} = \lambda g_{\mu\nu}$ for some constant λ (Deser, 1987). Hence, for an Einstein space the (generally curved) metric solves the full vacuum Einstein field equations (including a cosmological constant term which effectively acts as a source such that the energy-momentum tensor is proportional to the metric). Einstein manifolds with $\lambda = 0$ are called *Ricci-flat*. Note that a Ricci-flat manifold (i.e., $R_{\mu\nu} = 0$) is generally not flat with respect to curvature (i.e., generally $R^{\mu}_{\nu\sigma\rho} \neq 0$). The flat manifold with its flat metric η used throughout this chapter is flat in the strictest sense: *all* components of the curvature tensor vanish everywhere.

25 A spacetime $\langle M, g_{\mu\nu}\rangle$ is called *globally hyperbolic*, if it has a *Cauchy surface* Σ (Wald, 1984, p. 201), which is a three-dimensional spacelike hypersurface of M that defines a "slice of equal time". "[I]n a globally hyperbolic spacetime, the entire future and past history of the universe can be predicted (or retrodicted) from conditions at the instant of time represented by Σ. Conversely, in a non-globally hyperbolic spacetime we have a breakdown of predictability in the sense that a complete knowledge of conditions at a single 'instant of time' can never suffice to determine the entire history of the universe" (Wald, 1984, pp. 201–202). Notice that, under certain conditions, a number of theorems proved for globally hyperbolic spacetimes also apply to regions of non-globally hyperbolic spacetimes (Wald, 1984, p. 202).

26 I thank Brian Pitts for helpful discussion.

27 See Kment (2021) for an introduction.

28 Notably, the Minkowski metric can still be part of the overall ontology as discussed in Section 3.5.2 – it can still exist, albeit derivatively. In the context of spin-2 theory, this seems to mean that there are *two* non-fundamental metric fields, g and η. I shall later argue against this view.

29 Depending on one's general metaphysical stance with respect to the existence of derivative objects, one can still view g as part of the overall ontology, namely as a derivative object.

30 According to Lehmkuhl, "a mathematical object has gravitational significance if it plays a role in describing and explaining the phenomena we count as gravitational" (Lehmkuhl, 2008, p. 89).

31 Pooley himself, of course, concurs with the general sentiment that relationalism is not feasible for general relativity due to the g field's being dynamical and hence, according to the standard dynamical view, fundamental.

32 This may even help to downgrade the mentioned problems with the retention of the correct signature of g, and the correct relation of the respective null cones of η and g (Penrose, 1980; Pitts & Schieve, 2001a; 2001b) to technicalities, albeit highly non-trivial ones.

33 In string theory, it is the consistency requirement of the string worldsheet metric's conformal invariance which guarantees that the effective general-relativistic metric field g and all matter fields are Lorentz-invariant. Unlike in general relativity, "in perturbative string theory ... one is not free to choose matter field couplings; rather, the dynamical equations governing the behaviour of matter

fields on target space are an ineluctable consequence of the formalism" (Read, 2019, p. 107). Accordingly, both MR1 and MR2 are explained in string theory.

34 For a proper discussion of this issue, the reader is referred to Belot (2011) and especially Read (2016).

35 QGA takes η as ontologically independent. Therefore, the only possible explanation of MR2 – reducing the metric to matter field dynamics – is not available.

36 Fundamentally, the geometrical approach to spin-2 theory postulates the Minkowski metric and the matter fields (including h), whereas the dynamical approach postulates only the matter fields (including h).

37 See also Le Bihan and Linnemann (2019) who similarly argue that what they call the "explanatory gap" between general relativity and theories of quantum gravity is narrower than usually assumed.

38 One may of course argue that the spin-2 view is empirically preferable, as it arguably is more restrictive with respect to topology or causality. However, it is important to note that some key features of general relativity (e.g., the singularity theorems) might push against such assertions.

39 Hüttemann (2004) brings forward similar arguments to undermine the metaphysical priority of the micro level over the macro level in what he dubs "microphysicalism", i.e., the doctrine that it is the constituent parts that determine the whole.

40 Standardly, dependence is taken to be an asymmetric relation. I adopt this. Note, though, that Barnes (2018) takes dependence as symmetric.

41 As mentioned, the issue is even worse, since one could also use a different background, i.e., one could "split" g differently – which just again raises charges of conventionalism. Now, both camps could argue that the choice of η provides the best systematisation of the empirical data. As discussed, this is most convincing for a dynamical reading of η. On a geometrical view, the choice of one particular background, namely η, brings in both miracles – without explanation. But on a dynamical view, the choice of η just *is* the first miracle – which, of course, is itself unexplained: it simply happens to be the case that all matter fields exhibit Lorentzian dynamics in our world. There is a sense in which this is an example of reasoning via a continuity condition (see Section 6.3).

42 Arguably, all parts in a part–whole explanation are required to be pairwise ontologically independent. It seems odd to allow for redundant explanations like "the table ontologically depends on its legs and its atoms".

43 Specifically, η ontologically depends on h in *the same way* it ontologically depends on the ordinary matter fields: it depends on their symmetry properties (which are all Lorentzian according to MR1). The fundamentality of the ordinary matter fields (whatever they are) is uncontroversial in the geometrical–dynamical debate, and h is then just another matter field.

References

Álvarez, E. (1989). Quantum gravity: An introduction to some recent results. *Reviews of Modern Physics, 61(3)*, 561–604. doi:10.1103/RevModPhys.61.561

Barceló, C., Carballo-Rubio, R., & Garay, L. J. (2014). Unimodular gravity and general relativity from graviton self-interactions. *Physical Review D 89(12)*, 124019. doi:10.1103/PhysRevD.89.124019

Barnes, E. (2018). Symmetric Dependence. In R. Bliss, & G. Priest (Eds.), *Reality and Its Structure: Essays in Fundamentality* (pp. 50–69). Oxford: Oxford University Press.

Bartels, A. (2002). Das Gödel-Universum und die Philosophie der Zeit. In B. Buldt, et al. (Eds.), *Kurt Gödel. Wahrheit und Beweisbarkeit (Bd. II)* (pp. 231–250). Wien: öbv et htp-Verlag.

Belot, G. (2011). Background-independence. *General Relativity and Gravitation, 43(10)*, 2865–2884. doi:10.1007/s10714-011-1210-x

Boulware, D. G., & Deser, S. (1975). Classical general relativity derived from quantum gravity. *Annals of Physics, 89*, 193–240. doi:10.1016/0003-4916(75)90302-4

Brown, H. R. (2005). *Physical Relativity: Space-Time Structure from a Dynamical Perspective.* Oxford: Oxford University Press.

Butcher, L. M., Hobson, M., & Lasenby, A. (2009). Bootstrapping gravity: A consistent approach to energy-momentum self-coupling. *Physical Review D 80(8)*, 084014. doi:10.1103/PhysRevD.80.084014

Carlip, S. (2014). Challenges for emergent gravity. *Studies in History and Philosophy of Science Part B: Studies in History and Philosophy of Modern Physics, 46*, 200–208. doi:10.1016/j.shpsb.2012.11.002

Carroll, S. (2004). *Spacetime and Geometry: An Introduction to General Relativity.* San Francisco, CA: Addison Wesley.

Cheng, T.-P., & Li, L.-F. (1982). *Gauge Theory of Elementary Particle Physics.* Oxford: Clarendon Press.

Collins, J., Perez, A., Sudarsky, D., Urrutia, L., & Vucetich, H. (2004). Lorentz invariance and quantum gravity: An additional fine-tuning problem? *Physical Review Letters, 93(19)*, 191301. doi:10.1103/PhysRevLett.93.191301

Curiel, E. (2017). A Primer on Energy Conditions. In D. Lehmkuhl, G. Schiemann, & E. Scholz (Eds.), *Towards a Theory of Spacetime Theories. Einstein Studies, Vol. 13* (pp. 43–104). New, York, NY: Birkhäuser. doi:10.1007/978-1-4939-3210-8_3

Darrigol, O. (2007). A Faradayan principle for selecting classical field theories. *International Studies in the Philosophy of Science, 21(1)*, 35–55. doi:10.1080/02698590701305768

Darrigol, O. (2014). *Physics and Necessity: Rationalist Pursuits from the Cartesian Past to the Quantum Present.* Oxford: Oxford University Press. doi:10.1080/02698590701305768

Dautcourt, G., & Abdel-Megied, M. (2006). Revisiting the light cone of the Gödel universe. *Classical and Quantum Gravity, 23(4)*, 1269–1288. doi:10.1088/0264-9381/23/4/013

Deser, S. (1970). Self-interaction and gauge invariance. *General Relativity and Gravitation, 1*, 9–18. doi:10.1007/BF00759198

Deser, S. (1987). Gravity from self-interaction in a curved background. *Classical and Quantum Gravity, 4(4)*, L99–L105. doi:10.1088/0264-9381/4/4/006

Deser, S. (2010). Gravity from self-interaction redux. *General Relativity and Gravitation, 42*, 641–646. doi:10.1007/s10714-009-0912-9

Deser, S., & Henneaux, M. (2007). A note on spin-2 fields in curved backgrounds. *Classical and Quantum Gravity, 24*, 1683–1685. doi:10.1088/0264-9381/24/6/N01

Fang, J., & Frønsdal, C. (1979). Deformations of gauge groups. Gravitation. *Journal of Mathematical Physics, 20*, 2264–2271. doi:10.1063/1.524007

Feynman, R., Morinigo, F. B., Wagner, W. G., & Hatfield, B. (1995). *Feynman Lectures on Gravitation.* Reading, MA: Addison-Wesley.

Fierz, M., & Pauli, W. (1939). Relativistic wave equations for particles of arbitrary spin in an electromagnetic field. *Proceedings of the Royal Society of London A, 173*, 211–232. doi:10.1098/rspa.1939.0140

Friebe, C. (2012). Twins' paradox and closed timelike curves: The role of proper time and the presentist view on spacetime. *Journal for General Philosophy of Science, 43*, 313–326. doi:10.1007/s10838-012-9194-0

Gödel, K. (1949a). A Remark About the Relationship between Relativity Theory and Idealistic Philosophy. In K. Gödel (Ed.), *Collected Works, Vol. II: Publications 1938–1974* (pp. 202–207). New York, NY/Oxford: Oxford University Press, 1990.

Gödel, K. (1949b). An example of a new type of cosmological solutions of Einstein's field equations of gravitation. *Reviews of Modern Physics, 21*(3), 447–450. doi:10.1103/RevModPhys.21.447

Gupta, S. N. (1952a). Quantization of Einstein's gravitational field: General treatment. *Proceedings of the Physical Society. Section A, 65(8)*, 608–619. doi:10.1088/0370-1298/65/8/304

Gupta, S. N. (1952b). Quantization of Einstein's gravitational field: Linear approximation. *Proceedings of the Physical Society. Section A, 65(3)*, 161–169. doi:10.1088/0370-1298/65/3/301

Gupta, S. N. (1954). Gravitation and electromagnetism. *Physical Review, 96(6)*, 1683–1685. doi:10.1103/PhysRev.96.1683

Gupta, S. N. (1957). Einstein's and other theories of gravitation. *Reviews of Modern Physics, 29(3)*, 334–336. doi:10.1103/RevModPhys.29.334

Hossenfelder, S. (2018). *Lost in Math: How Beauty Leads Physics Astray.* New York, NY: Basic Books.

Huggins, E. R. (1962). *Quantum mechanics of the interaction of gravity with electrons: theory of a spin-two field coupled to energy.* Ph.D. dissertation, California Institute of Technology. doi:10.7907/05S4-6910

Hüttemann, A. (2004). *What's Wrong with Microphysicalism?* London/New York, NY: Routledge.

Kment, B. (2021). Varieties of Modality. In E. N. Zalta (Ed.), *The Stanford Encyclopedia of Philosophy* (Spring 2021 ed.). https://plato.stanford.edu/archives/spr2021/entries/modality-varieties

Kraichnan, R. H. (1955). Special relativistic derivation of generally covariant gravitation theory. *Physical Review, 98(4)*, 1118–1122. doi:10.1103/PhysRev.98.1118

Kraichnan, R. H. (1956). Possibility of unequal gravitational and inertial masses. *Physical Review, 101(1)*, 482–488. doi:10.1103/PhysRev.101.482

Kupersztych, J. (1988). Is there a Link between Gauge Invariance, Relativistic Invariance and Electron Spin? In M. E. Noz, & Y. S. Kim (Eds.), *Special Relativity and Quantum Theory. A Collection of Papers on the Poincaré Group* (pp. 447–457). Dordrecht: Kluwer Academic Publishers.

Le Bihan, B., & Linnemann, N. (2019). Have we lost spacetime on the way? Narrowing the gap between general relativity and quantum gravity. *Studies in History and Philosophy of Science Part B: Studies in History and Philosophy of Modern Physics, 65*, 112–121. doi:10.1016/j.shpsb.2018.10.010

Lehmkuhl, D. (2008). Is Spacetime a Gravitational Field? In D. Dieks (Ed.), *The Ontology of Spacetime Vol. II. Philosophy and Foundations of Physics, Vol. 4* (pp. 83–110). Amsterdam: Elsevier.

Linnemann, N., Smeenk, C., & Baker, M. R. (2023). GR as a classical spin-2 theory? *Philosophy of Science*, 1–19. doi:10.1017/psa.2023.56

Maggiore, M. (2008). *Gravitational Waves. Volume 1: Theory and Experiment.* Oxford: Oxford University Press.

Martens, N. C. (2019). The metaphysics of emergent spacetime theories. *Philosophy Compass, 14(7)*, e12596. doi:10.1111/phc3.12596

Massimi, M., & Redhead, M. (2003). Weinberg's proof of the spin-statistics theorem. *Studies in History and Philosophy of Science Part B: Studies in History and Philosophy of Modern Physics, 34(4)*, 621–650. doi:10.1016/S1355-2198(03)00066-2

Mattingly, D. (2005). Modern tests of Lorentz invariance. *Living Reviews in Relativity*, 8, 5. doi:10.12942/lrr-2005-5

Maudlin, T. (1993). Buckets of water and waves of space: Why spacetime is probably a substance. *Philosophy of Science*, 60, 183–203. doi:10.1086/289728

Misner, C. W., Thorne, K. S., & Wheeler, J. A. (1973). *Gravitation.* Princeton, NJ: Princeton University Press.

Noz, M. E., & Kim, Y. S. (1986). *Theory and Applications of the Poincare Group.* Dordrecht: D. Reidel Publishing Company.

Ogievetsky, V. I., & Polubarinov, I. V. (1965). Interacting field of spin-2 and the Einstein equations. *Annals of Physics, 35(2)*, 167–208. doi:10.1016/0003-4916(65)90077-1

Ortín, T. (2015). *Gravity and Strings.* Cambridge: Cambridge University Press.

Ortín, T. (2017). Higher order gravities and the strong equivalence principle. *Journal of High Energy Physics, 152*, 1–14. doi:10.1007/JHEP09(2017)152

Padmanabhan, T. (2003). Cosmological constant – The weight of the vacuum. *Physics Reports, 380(5–6)*, 235–320. doi:10.1016/S0370-1573(03)00120-0

Padmanabhan, T. (2008). From gravitons to gravity: Myths and reality. *International Journal of Modern Physics D, 17(03n04)*, 367–398. doi:10.1142/S0218271808012085

Papapetrou, A. (1948). Einstein's theory of gravitation and flat space. *Proceedings of the Royal Irish Academy. Section A: Mathematical and Physical Sciences, 52*, 11–23.

Penrose, R. (1980). On Schwarzschild Causality. A Problem for "Lorentz Covariant" General Relativity. In F. J. Tipler (Ed.), *Essays in General Relativity. A Festschrift for Abraham Taub* (pp. 1–12). New York, NY: Academic.

Pitts, J. B. (2016). Einstein's equations for spin 2 mass 0 from Noether's converse Hilbertian assertion. *Studies in History and Philosophy of Science Part B: Studies in History and Philosophy of Modern Physics, 56*, 60–69. doi:10.1016/j.shpsb.2016.08.008

Pitts, J. B. (2017). Progress and Gravity: Overcoming Divisions Between General Relativity and Particle Physics and Between Physics and HPS. In K. Chamcham, J. Silk, J. D. Barrow, & S. Saunders (Eds.), *The Philosophy of Cosmology* (pp. 263–282). Cambridge: Cambridge University Press.

Pitts, J. B., & Schieve, W. C. (2001a). Light cone consistency in bimetric general relativity. *AIP Conference Proceedings, 586(1)*, 763–765. doi:10.1063/1.1419652

Pitts, J. B., & Schieve, W. C. (2001b). Null Cones in Lorentz-Covariant General Relativity. *Null Cones in Lorentz-Covariant General Relativity*. http://philsci-archive.pitt.edu/481/

Pitts, J. B., & Schieve, W. C. (2001c). Slightly bimetric gravitation. *General Relativity and Gravitation, 33*, 1319–1350. doi:10.1023/A:1012005508094

Pitts, J. B., & Schieve, W. C. (2007). Universally coupled massive gravity. *Theoretical and Mathematical Physics, 151*, 700–717. doi:10.1007/s11232-007-0055-7

Poincaré, H. (1905). *Science and Hypothesis*. London/Newcastle-on-Tyne: The Walter Scott Publishing Co.

Pooley, O. (2013). Substantivalist and Relationalist Approaches to Spacetime. In R. Batterman (Ed.), *The Oxford Handbook of Philosophy of Physics* (pp. 522–586). Oxford: Oxford University Press. http://philsci-archive.pitt.edu/9055/

Pooley, O. (2017). Background Independence, Diffeomorphism Invariance, and the Meaning of Coordinates. In D. Lehmkuhl, G. Schiemann, & E. Scholz (Eds.), *Towards a Theory of Spacetime Theories. Einstein Studies, Vol. 13* (pp. 105–143). New York, NY: Birkhäuser. doi:10.1007/978-1-4939-3210-8_4

Preskill, J., & Thorne, K. S. (1995). Foreword. In R. P. Feynman, F. B. Morinigo, W. G. Wagner, & B. Hatfield (Eds.), *Feynman Lectures on Gravitation* (pp. vii–xxx). Reading, MA: Addison-Wesley.

Read, J. (2016). Background Independence in Classical and Quantum Gravity. https://ora.ox.ac.uk/objects/uuid:b22844af-aac9-4adc-a6c5-1e2815c59655

Read, J. (2019). On miracles and spacetime. *Studies in History and Philosophy of Science Part B: Studies in History and Philosophy of Modern Physics, 65*, 103–111. doi:10.1016/j.shpsb.2018.10.002

Read, J. (2020). Explanation, Geometry, and Conspiracy in Relativity Theory. In C. Beisbart, T. Sauer, & C. Wüthrich (Eds.), *Thinking About Space and Time: 100 Years of Applying and Interpreting General Relativity. Einstein Studies, Vol. 15* (pp. 173–205). Basel: Birkhäuser.

Read, J., Brown, H. R., & Lehmkuhl, D. (2018). Two miracles of general relativity. *Studies in History and Philosophy of Science Part B: Studies in History and Philosophy of Modern Physics, 64*, 14–25. doi:10.1016/j.shpsb.2018.03.001

Rosen, N. (1940a). General relativity and flat space. I. *Physical Review, 57*, 147–150. doi:10.1103/PhysRev.57.147

Rosen, N. (1940b). General relativity and flat space. II. *Physical Review, 57*, 150–153. doi:10.1103/PhysRev.57.150

Salimkhani, K. (2020). The dynamical approach to spin-2 gravity. *Studies in History and Philosophy of Science Part B: Studies in History and Philosophy of Modern Physics, 72*, 29–45. doi:10.1016/j.shpsb.2020.05.002

Sexl, R. U., & Urbantke, H. K. (2002). *Gravitation und Kosmologie*. Heidelberg: Spektrum.

Shankar, R. (2008). *Principles of Quantum Mechanics* (2nd ed.). Springer Science.

Smeenk, C. (2013). Philosophy of Cosmology. In R. Batterman (Ed.), *Oxford Handbook of Philosophy of Physics* (pp. 607–652). Oxford: Oxford University Press.

Streater, R. F., & Wightman, A. S. (2000). *PCT, Spin and Statistics, and All That* (5th ed.). Princeton, NJ: Princeton University Press.

Thirring, W. E. (1959). Lorentz-invariante Gravitationstheorien. *Fortschritte der Physik, 7(2)*, 79–101. doi:10.1002/prop.19590070202

Thirring, W. E. (1961). An alternative approach to the theory of gravitation. *Annals of Physics, 16(1)*, 96–117. doi:10.1016/0003-4916(61)90182-8

Thorne, K. S. (1993). Closed timelike curves. In *13th Conference on General Relativity and Gravitation (GR-13)* (pp. 295–315).

Tung, W.-K. (2003). *Group Theory in Physics. An Introduction to Symmetry Principles, Group Representations, and Special Functions in Classical and Quantum Physics*. London: World Scientific Publishing.

Wald, R. M. (1984). *General Relativity*. Chicago, IL: The University of Chicago Press.

Wald, R. M. (1986). Spin-two fields and general covariance. *Physical Review D, 33*, 3613–3625. doi:10.1103/PhysRevD.33.3613

Weinberg, S. (1964a). Derivation of gauge invariance and the equivalence principle from Lorentz invariance of the S-matrix. *Physics Letters, 9(4)*, 357–359. doi:10.1016/0031-9163(64)90396-8

Weinberg, S. (1964b). Photons and gravitons in S-matrix theory: Derivation of charge conservation and equality of gravitational and inertial mass. *Physics Review, 135(4B)*, B1049–B1056. doi:10.1103/PhysRev.135.B1049

Weinberg, S. (1965). The Quantum Theory of Massless Particles. In S. Deser, & K. Ford (Eds.), *Lectures on Particles and Field Theory* (pp. 405–485). Englewood Cliffs, NJ: Prentice-Hall.

Weinberg, S. (1972). *Gravitation and Cosmology: Principles and Applications of the General Theory of Relativity*. New York: Wiley.

Weinberg, S. (1995). *The Quantum Theory of Fields. Volume I: Foundations*. Cambridge: Cambridge University Press.

Weinberg, S. (1999). What Is Quantum Field Theory, and What Did We Think It Was? In T. Y. Cao (Ed.), *Conceptual Foundations of Quantum Field Theory* (pp. 241–251). Cambridge: Cambridge University Press.

Wyss, W. (1965). Zur Unizität der Gravitationstheorie. *Helvetica Physica Acta, 38(5)*, 469–480. doi:10.5169/seals-113605

Yourgrau, P. (1999). *Gödel Meets Einstein. Time Travel in the Gödel Universe*. La Salle: Open Court Publishing Co.

5 Spacetime in Quantum Gravity

5.1 The Quest for a Theory of Quantum Gravity

5.1.1 What Is Quantum Gravity?

Infamously, there is no consensus on what the best approach is to finding the correct theory of quantum gravity.[1] In a sense, this is largely because researchers disagree on what exactly the question is in the first place:

> [n]o question about quantum gravity is more difficult than the question, 'What is the question?'
>
> (Wheeler, 1984, p. 224)

The majority of philosophical works on the subject take it that what goes by the label "quantum gravity" is supposed to denote some future theory that consistently brings together (in a sense yet to be spelled out) what is often considered the divided basis of modern physics: quantum theory and general relativity.[2] Typically, such a theory is assumed to successfully apply (in a sense yet to be spelled out) the lessons from quantum theory (yet to be specified) to gravity or spacetime, depending on one's methodology and one's interpretation of general relativity. This is why the terms "quantisation of gravity" and "quantisation of spacetime" are frequently used synonymously for the research programme. Relatedly, since a quantisation of spacetime seems to suggest that (classical) spacetime "emerges" from an underlying quantum structure, the notion "emergence of spacetime" – or, conversely, "disappearance of spacetime" – is ubiquitous in the debate.

In fact, it is fair to say that a common perspective on the issue does assume "not only that there is no fundamental spacetime substance, but also that spatiotemporal relations are not fundamental" (Esfeld, 2021, p. S355):

> Thus, the claim is not that, as there is a shift from Euclidean to Riemannian geometry in the transition from Newtonian gravitation to Einstein's general theory of relativity, so there may be another such

DOI: 10.4324/9781003404149-5

shift in geometry in the transition to a quantum theory of gravity. The claim is much more radical, namely that after that transition, there will be nothing left in fundamental physics that is like the spacetime or the spatiotemporal relations with which we are familiar.

(Esfeld, 2021, pp. S355–S356)

It is important to note, though – and Esfeld points this out himself – that not all theories that quantise gravity do also explicitly quantise spacetime,[3] or do away with spatiotemporal structure altogether (e.g., quantum-field-theoretic approaches and string theory).

Due to the various specifications needed and due to different methods used (regarding quantisation, for example), these slogans generally refer to vastly different theory proposals with respect to aims, research questions, and conclusions. For instance, approaches that seek to quantise spacetime will typically explore how classical spacetime can be obtained from quantum structure, while approaches that intend to give, first and foremost, a quantum-theoretical description of gravity do not need to worry so much about the theory having the correct classical limit (general relativity), but will typically focus on issues like UV completion – i.e., the problem that quantum theories of gravity are divergent and, hence, non-predictable at high energies (the ultraviolet limit).[4]

Another scheme that is often adopted (see Kiefer, 2007) divides the theory space into so-called *primary theories of quantum gravity* that attempt to quantise general relativity (according to the different standard methods, like canonical or covariant quantisation), and *secondary theories of quantum gravity* that attempt to work out a fundamental quantum theory (e.g., string theory) that contains – amongst others – quantum gravity in some limit. Note that this classification is based on how the approaches proceed, i.e., applies to investigating the context of discovery.[5] Systematically, i.e., with respect to the context of justification (with which I shall mostly be concerned here), the respective approaches may nonetheless be related. For instance, Weinberg (1999) emphasises the relation between quantum-field-theoretic, i.e., covariant, approaches and string theory.

The slogans "quantisation of spacetime", "emergence of spacetime", or "disappearance of spacetime" already indicate that theories of quantum gravity might offer new insights into the issue of the fundamentality of spacetime or spatiotemporal properties. Philosophically, the issue of spacetime possibly being a non-fundamental structure (in some sense yet to be specified) due to quantum theories of gravity is taken very seriously in both philosophy of physics[6] and mainstream metaphysics.[7] It has even given rise to the formation of a new research field, the metaphysics of quantum gravity.[8] But why does physics pursue the research programme in the first place?

5.1.2   *Why Quantum Gravity?*

Amongst the standard arguments for why there is a need for a theory of quantum gravity is the prediction of certain phenomena, like Hawking radiation. Roughly speaking, the description of such phenomena requires both general relativity and quantum theory, since it involves strong gravitational fields and quantum mechanical effects. More precisely, certain features of gravity, like the scale dependence of the gravitational coupling constant,[9] are standardly argued to show that gravitational effects cannot be neglected at the Planck scale where gravity dominates all other forces.[10]

In addition, many appeal to problems in the standard model of cosmology (regarding the initial state of the universe at the "big bang"), black hole evolution, consistency issues in quantum field theory, conceptual issues regarding, for example, the operational accessibility of spacetime or the alleged inconsistency of the different notions of time (in quantum theory and general relativity), or point to a lack of unification in fundamental physics.[11] Essentially, the reasoning is that these issues indicate some sort of theoretical incompleteness of present-day physics. The majority of physicists and philosophers accept these arguments as conclusive and accept that the research programme of quantum gravity is well justified.

Some, however, contest that such arguments can make a strong case in favour of the general research programme (e.g., Mattingly, 2005 and, certainly to a lesser degree, Wüthrich, 2005). This is mostly because contestable philosophical reasons, not rigorous ones derived from physics, are argued to play a decisive role in justifying the research programme; and, because, strictly speaking, genuine quantum-gravitational phenomena have not yet been observed, but are only *expected* to occur at some remote energy scale, namely the Planck scale:

> Quantum gravity presents something of a unique puzzle for the philosophy of science. For in a very real sense, there is no such thing as quantum gravity. Despite near unanimous agreement among physicists that a quantum theory of gravitation is needed to reconcile the contradictions between general relativity and quantum mechanics, there are no pressing empirical issues that require this resolution – the regime in which one would expect to observe a conflict between the claims of general relativity and quantum mechanics is at the Planck scale. Thus the question naturally arises "Why quantize gravity?"
>
> (Mattingly, 2005, p. 325)

The critic's verdict is that there is no empirical need whatsoever to construct a theory that quantises gravity (or spacetime). There is a sense in which this is correct. Quantum mechanics, or rather the standard model of particle physics, and general relativity are in perfect agreement with all available, and – concerning quantum-gravitational effects – presumably even all expectable data.[12] In addition, all known methods for promoting a classical theory to a quantum theory "run into difficulties when applied to general relativity" (Wald, 1984, p. 380).

According to the critic, this is not to say that present physics theories are, in fact, perfectly fine, and that there is nothing to do for physics at all. As it stands, general relativity is a fully classical theory and, hence, does have its problems: general relativity treats matter fields classically – contrary to what we already know. But even though general relativity is wrong about matter, it might still be right about spacetime and gravitation, or so opponents of quantum gravity argue[13]:

> There is clearly *something* wrong with the general relativistic treatment of matter fields as classical. Very well. Let us stipulate that an acceptable theory of gravitation will take due note of the quantum nature of the fields to which it couples.
>
> (Mattingly, 2005, p. 325)

Consequently, the critic advises to not seek a full theory of quantum gravity that quantises gravity or potentially does away with spacetime altogether, but simply find a way to introduce quantum matter to general relativity. Such an approach leaves spacetime and gravity as they are (or almost as they are). In this sense, "consistently bringing together quantum theory and general relativity" receives what one could call a "minimal reading". It is here where so-called *semi-classical theories* (see Carlip, 2008) take their departure and

> supply the wedge that Callender and Huggett [Huggett and Callender (2001b)] drive between the two disparate questions of whether we need a quantum theory of gravity and of whether such a need implies the quantization of gravity.
>
> (Wüthrich, 2005, p. 780)

The advocate of a semi-classical theory regards all approaches which go beyond remedying the specific shortcoming that general relativity ignores the fact that matter fields are quantum as physically unwarranted. From their vantage point, all such approaches rely exclusively on (hardly convincing) philosophical arguments.

Arguably, such reasoning is fuelled by a geometrical interpretation of general relativity and some form of fundamentalism about spacetime. For example, proponents of a semi-classical theory are likely to agree with the standard geometrical reading that "according to GTR [the general theory of relativity], gravity simply is not a force" (Maudlin, 1996, p. 143), unlike the electromagnetic, the weak, and the strong interaction – and, therefore, should not be quantised. In other words, the idea is that gravity merely determines the structure of the "arena", i.e., spacetime itself, in which all the other fields propagate and interact (Álvarez, 1989, p. 561). Hence, spacetime (or the gravitational field) is deemed to be fundamentally distinguished from and unaffected by the quantum description of matter.

To see why the semi-classical approach is typically considered untenable nevertheless and, as a consequence, a full theory of quantum gravity is considered crucial, let me take a closer look at the semi-classical proposal and then discuss three counter arguments.

As mentioned, the general idea is to quantise neither spacetime nor gravity, but to simply modify the Einstein field equations to accommodate the fact that matter fields are fundamentally quantum, while maintaining that gravitation and spacetime are fundamentally classical, i.e., "non-quantum". At first, this does not seem to pose any conceptual difficulty, because general relativity does not specify anything about the microstructure of matter (e.g., Lehmkuhl, 2019). Originally, semi-classical theories have been proposed, amongst others by Møller (1962) and Rosenfeld (1963), the latter of whom argues that

> [t]he combination of a classical description of the gravitation field with a quantal description of the other fields need not present any difficulty, either of practical formulation or of principle. In formulating such a theory, one would naturally assume that the source of the gravitation field would be determined by the expectation value of the energy-momentum density operator.
>
> (Rosenfeld, 1963, p. 354)

So, technically, such a semi-classical description is achieved straightforwardly by replacing the classical energy-momentum tensor $T_{\mu\nu}$ by its expectation value $\langle T_{\mu\nu}\rangle$ in the Einstein field equations:

$$G_{\mu\nu} = 8\pi\langle T_{\mu\nu}\rangle \tag{5.1}$$

with the Einstein tensor $G_{\mu\nu} = R_{\mu\nu} - 1/2\, R\, g_{\mu\nu}$, which can be viewed to represent spacetime geometry. Note that the superposition principle for matter states is lost in Eq. (5.1); different matter states are associated with different spacetimes (Wald, 1984, p. 382).

Apart from the question of whether this theory is empirically adequate, the question is whether this theory is consistent – both in itself and with established theorems of physics. Regarding self-consistency, note that the expectation value of the energy-momentum tensor does not behave as a purely classical object – unlike the Einstein tensor $G_{\mu\nu}$ and despite its exhibiting some convenient properties according to Ehrenfest's theorem, which links the quantum-mechanical expectation value to classical equations of motion. This results in the theory's suffering from discontinuities as, for example, Wald (1984, pp. 382–383) points out – and hence a first argument against semi-classical theories can be constructed.

Here is Wald's version of the argument. Assume the semi-classical picture of Eq. (5.1) and consider a state in which, with probability 1/2, all the matter is located either in region O_1 or in another region O_2 disjoint from O_1. Then, Eq. (5.1) yields a metric field g for the following matter distribution: half the matter is in region O_1 and half the matter is in region O_2. Now one performs a position measurement to see where the matter actually is. According to the rules of quantum mechanics, one instantaneously obtains the following definite answer: either all the matter is in region O_1 or all the matter is in region O_2. With Eq. (5.1) still holding, this results in a discontinuous, acausal change of the metric field g. Therefore, or so Wald argues, semi-classical theories are untenable. Note, though, that proponents of semi-classical theories might be able to bring forward an argument to the effect that Wald's conclusion depends on a particular interpretation of quantum theory (see Mattingly, 2005; Wüthrich, 2005).

While Wald's argument tells against a specific semi-classical theory of gravitation and relies on the general assumption that physical theories should be (causally) well-behaved and smooth, a second argument has force against all kinds of semi-classical theories and relies on an empirically well-confirmed theorem of physics, namely Heisenberg's uncertainty principle – a cornerstone of quantum theory. Here is the argument. Assume that gravitational waves exist, as has recently been validated empirically (Abbott et al., 2016). If gravitational waves were classical, they would in principle allow for *arbitrarily precise* measurements. This means that we could in principle use gravitational waves to measure both the position *and* the velocity of any quantum object to arbitrary precision – which would violate the Heisenberg uncertainty principle.

Essentially, this is a crude version of the more careful argument by Eppley and Hannah (1977). In fact, ultimately the argument shows that any interaction between some *generic* (not necessarily gravitational) classical field and a quantum field is inconsistent – either with Heisenberg's uncertainty principle, momentum conservation, or Einstein causality (Eppley & Hannah, 1977) – and therefore forbidden. Accordingly, given that all the matter fields are fundamentally quantum, the argument suggests that gravity needs to be quantised as well. But again, caveats apply due to their assuming a collapse interpretation of quantum theory (Wüthrich, 2005, p. 780).[14]

A third and highly influential argument for why most physicists agree that there is a need for a theory of quantum gravity combines the specificity of the first argument on whether *gravitation and spacetime* can be classical and the generality of the second argument that aims to exclude *any* semi-classical proposal. The argument seeks to infer the non-fundamentality of spacetime from establishing a breakdown of an operationally sensible notion of classical spacetime.[15]

Essentially, the argument rests on the Heisenberg uncertainty principle and the fact that gravity couples to all forms of energy and momentum. When probing some object, say an electron, up to an uncertainty Δx (or accuracy $1/\Delta x$) in position, Heisenberg's uncertainty principle tells us that this is related to an uncertainty in the electron's momentum Δp by $\Delta x \cdot \Delta p \geq \hbar/2$. Probing an electron up to an uncertainty Δx in position essentially means to confine this electron to a spatiotemporal region of "radius" Δx. So, confining the electron in a spatiotemporal region of radius Δx is accompanied by a corresponding uncertainty in the electron's momentum of $\Delta p \propto 1/\Delta x$. In other words, as energy is proportional to momentum, a position measurement up to an uncertainty Δx deposits an amount of energy $\Delta E \propto 1/\Delta x$ in the spatiotemporal region of radius Δx. Hence, when probing to higher accuracy, i.e., when probing a smaller spatiotemporal region, we deposit more energy in this (smaller) region. Since gravity couples to all forms of energy and momentum,[16] this generates a gravitational field which gets stronger when increasing the accuracy, i.e., when decreasing the radius of the region. At some point (the Planck length $l_P \approx 10^{-33}$ cm), the spatiotemporal region is so small that so much energy is deposited in it, and so the gravitational field gets so strong, that the region collapses into a black hole. In other words, for $\Delta x = l_P$ the region's radius is smaller than its Schwarzschild radius r_s; the region's Schwarzschild radius increases as the energy deposited in the region increases when decreasing the radius of the region ($r_s \propto \Delta E \propto 1/\Delta x$).

This "Heisenberg collapse" of a sufficiently small spatiotemporal region is often taken to establish an operational notion of a "minimal length" (the Planck length l_P) and therefore blocks an operationally defined notion of arbitrarily short spatiotemporal distances – *contra* the conception of classical continuous spacetime (which naturally has a notion of arbitrarily short spatiotemporal distances)[17]:

> combining Heisenberg's uncertainty principle with Einstein's theory of classical gravity leads to the conclusion that ordinary spacetime loses any operational meaning in the small.
>
> (Doplicher et al., 1995, p. 189)[18]

There is simply no operational way of probing physics at Planckian spatiotemporal scales. Arkani-Hamed (2010b) takes this to indicate that the notion of classical continuous spacetime *as such* is not only operationally problematic, but *non-fundamental* – which is how I propose to read his use of the notion "existence":

> at this point you could take two attitudes. One of them is that space and time actually do exist. They're there. It's just we can't measure it. Everything we've learned in the history of physics says that that's a very reactionary attitude. And every quantity that we haven't managed to associate with something that can in principle be observed, that's telling us something very important. It's telling us that that quantity actually doesn't exist. There's no such thing as spacetime at arbitrarily short scales. Something has to give.
>
> (Arkani-Hamed, 2010b)

So, according to Arkani-Hamed, the Heisenberg collapse reveals spacetime as approximate – a non-fundamental structure that arises from something more fundamental.[19] Note that Arkani-Hamed explicitly rejects interpretations of the Heisenberg collapse that ascribe a literal ontological meaning to the minimal length à la "spacetime is fundamentally discrete". The reason is that a fundamentally discrete spacetime will generally violate Lorentz invariance; some, however, uphold that local Lorentz invariance can actually be retained in certain discrete spacetime scenarios.[20] Instead, Arkani-Hamed takes the Heisenberg collapse to indicate that the principles of physics responsible for the Heisenberg collapse, i.e., Lorentz invariance and unitarity (which are the core principles of special relativity and quantum mechanics, respectively), need to be reconciled in a more fundamental theory. Notably, Arkani-Hamed does not attribute the Heisenberg collapse to *general* relativity, or a principle thereof. This is because he views general relativity as deduced from a relativistic quantum spin-2 theory (which is based on special relativity and quantum mechanics).

Now, is it convincing to claim that the Heisenberg collapse implies that spacetime is non-fundamental (or, in light of Jaksland and Salimkhani (2023), that certain spatiotemporal aspects, like metrical aspects or continuity, are non-fundamental)? Let me focus on two points that are discussed in Martens (2019) and Wüthrich (2005), respectively.

First, Martens argues that Arkani-Hamed draws his ontological conclusions from an epistemological argument:

> when *probing* spacetime at small length scales we destroy it, which makes it impossible for us to *know* anything about spacetime at such scales. ... We cannot leap from the epistemological problem of not being able to know the local structure of spacetime to the ontological claim that there is no such thing ....
>
> (Martens, 2019, p. 6)

Martens considers the possibility to invoke Occam's razor ("do not postulate more entities than necessary") to "close the gap between epistemology and ontology" (Martens, 2019, p. 6):

> but it is simply not clear in this context that the non-emergentist [who takes spacetime as fundamental] is postulating any extra, unnecessary entities. Spacetime is one big entity; all scales come as a package deal. It seems to be a category mistake to treat the small scales as extra entities that we could just throw out. In fact, isn't it the emergentist [who takes spacetime to be non-fundamental] who is postulating new entities? Thus, I suggest we hold our horses and refrain from giving in to Arkani-Hamed's operationalist urge to conclude, merely from the epistemological argument that we destroy spacetime when probing it at small scales, that there never was any spacetime at those scales to begin with.
>
> (Martens, 2019, p. 6)

It strikes me that this reply is missing its mark. First of all, as Martens adds himself, there are ways to construct the Heisenberg collapse argument in a straightforwardly ontological fashion[21] by noticing that quantum fluctuations will – independently of epistemological concerns – collapse spatiotemporal regions into black holes at the Planck scale. After all, already in Arkani-Hamed's version of the argument, the work is done by the *conceptual* connection between length scale and energy due to Heisenberg's uncertainty principle and the fact that there is gravity, *not* by the operational or epistemological gloss.[22] It is this conceptual issue (presumably a consistency issue) which is the reason for there being no operational way to define spacetime at arbitrarily short scales – indeed, it is precisely the issue that such semi-classical theories are not UV-complete (an issue that applies to quantum-field-theoretic proposals as well).

One may also simply argue that Arkani-Hamed is right when pointing out that physics should – and usually does – insist on some notion of in-principle operational or epistemological access for ontologically committing to an entity, precisely because a lack thereof might reveal deeper conceptual issues; especially with respect to heuristics, it seems reasonable to take operational definitions seriously (see, for example, Einstein regarding the relativity of simultaneity and Brown regarding the dynamical approach to spacetime). Philosophically, similar arguments have been advanced, for example by Leibniz, to argue for relationalism and against substantivalism of spacetime (see Section 2.2.1).

Secondly, spacetime does indeed come as what Martens calls a "package deal". It is part of the very essence of the notion of classical spacetime that it applies to all scales. But this is precisely why the appearance of conceptual problems at some scale calls into question the notion as such when it comes to questions of fundamentality – especially when it is about a scale that is usually considered particularly relevant for questions of fundamentality. I agree with Martens that invoking Occam's razor does not help – but why should this bother Arkani-Hamed? His argument does not treat small-scale spacetime as an extra (i.e., unnecessary and redundant) entity, but seeks to establish that the whole package, spacetime as such, is not empirically adequate (for the time being, in a broad sense of the notion including thought experiments) at what he takes to be the fundamental scale that determines the fundamental ontology.

Applying Occam's razor would constitute a different argument that Arkani-Hamed could bring forward. He could argue that, if it is relevant at all, Occam's razor needs to be applied to *large-scale* spacetime. Such an argument would insist that at large scales the notion might be empirically adequate, but is still redundant to another, more fundamental description. Arkani-Hamed would then need to say what replaces spacetime fundamentalism at large scales (similarly to what I put forward in this book).[23] At small scales where the notion breaks down due to the Heisenberg collapse, Occam's razor has nothing to contribute – the notion is already dismissed for not being empirically adequate. Again, as soon as spacetime's fundamentality has been called into question at some scale, the notion as such is dismissed due to its being a package deal.

Thirdly, Martens misses a crucial part of the argument. Arkani-Hamed pleads for a breakdown of spacetime at *arbitrarily* short scales:

> What if you say damn, that's not what I wanted to happen? Let me build an even higher energy accelerator. Let me build an even more powerful microscope. What would happen? You make an even bigger black hole! It gets even worse. … you can't even know what's going on at 10^{-31} cm anymore [which in principle you can, since 10^{-31} cm < l_P; my remark]. That's being collapsed into a black hole.
>
> (Arkani-Hamed, 2010b)

When subtracting the epistemological gloss (e.g., by rephrasing the argument in straightforwardly ontological terms), the key issue becomes apparent: the concept of classical spacetime implies a notion of arbitrarily short distances which is accompanied by a Heisenberg collapse at *arbitrarily large* scales. Hence, the Heisenberg collapse argument does not only establish a minimal length scale of $l_P \approx 10^{-33}$ cm (due to the Heisenberg collapse of spatiotemporal regions of Planck length radii), but it straightforwardly establishes a conceptual breakdown of the whole package of spacetime altogether since the concept of spatiotemporal regions of *arbitrarily small* radii – in particular, of $l < l_P$ – shifts the minimal length to arbitrarily large scales $l > l_P$. Recall that the region's Schwarzschild radius increases as the energy deposited in the region increases for decreasing the radius of the region ($r_s \propto \Delta E \propto 1/\Delta x$) – for arbitrarily small Δx, r_s becomes arbitrarily big. Note that this argument does not appeal to the presupposition that spacetime is a package deal, but *demonstrates* that spacetime is a package deal. This is why Arkani-Hamed takes – or should take – the Heisenberg collapse seriously and I agree that this does provide a conceptual hint that spacetime is non-fundamental; or, put more carefully, that the metrical aspects of spacetime are non-fundamental.

But why is all of this telling against the fundamentality of spatiotemporal structure (arguably, metric structure specifically) *in the metaphysical sense of the notion, namely ontological independence*? At first sight, it seems perfectly fine to accept some structure as ontologically independent but operationally inaccessible or conceptually inconsistent. So, have we tacitly moved to a different (operational) notion of fundamentality here? Or, have we tacitly assumed that fundamentals additionally need to be conceptually consistent?[24] This issue arguably requires further scrutiny, but here is how one could connect the presentation above to the ontological independence notion of fundamentality. In short, arbitrarily short distances are operationally inaccessible and conceptually inconsistent, *because* metrical aspects are *not* ontologically independent. At least some metrical aspects (e.g., continuity) depend on the presence (or absence) of something else: whether the spacetime metric of some world can be surveyed by material rods and clocks to arbitrary precision depends on whether there is matter, gravity, and the Heisenberg uncertainty principle in this world. After all, there are arbitrary short distances in classical spacetimes *in virtue of* the absence of the Heisenberg principle. Similarly, in a world without gravity, but with the Heisenberg uncertainty principle, say, a purely quantum-electrodynamical world, there are arbitrary short distances *in virtue of* the absence of gravity.[25]

Let us dwell on this a bit further. Take a window and a stone. Does the window ontologically depend on the stone just because the stone can shatter the window? Presumably, most will disagree. A world with a stone and an intact window is perfectly consistent. The stone's shattering the window is a mere possibility. Whether the stone actually shatters the window depends on various boundary conditions. It is not a universal law of nature that as soon as a world contains a window and a stone, that stone will shatter the window. Thus, the window does not exist in virtue of the absence of the stone, both are ontologically independent. In the spacetime case, however, there are such laws: as soon as matter fields, gravity and the Heisenberg uncertainty principle are around, quantum fluctuations *ensure* the breakdown of arbitrarily short distances and continuous spacetime.

Let me move forward to Wüthrich's (2005) objection. Wüthrich raises doubts about the Heisenberg collapse argument's ability to establish a specific ontological conclusion. Based on his assessment of Doplicher et al.'s (1995) version of the argument, Wüthrich insists that

> one cannot expect that all principles that underpin the above premises [of the Heisenberg collapse argument] will still be valid in full quantum gravity. But if not all premises hold, then the argument will of course collapse. Because the source energy associated with $T_{\mu\nu}$ is quantum and the argument as given is thus strictly semi-classical, the objection could conclude, it may very well turn out that the discreteness emerges only as an artefact of the manner in which quantum mechanics and general relativity were combined at the semi-classical level. Therefore, the argument as given so far must be complemented by a second part asserting that the operationally discrete spacetime at the semi-classical level results from an underlying discreteness at the fundamental Planck level.
>
> Unfortunately, Doplicher and his collaborators do not provide such an addition. But neglecting to take this second leg seriously amounts to begging the question. It is undoubtedly true that if spacetime is discrete at the Planck level, it is reasonable to expect some signatures of this discreteness to surface at the semi-classical level. But the converse is not true, exactly because some or all of the premises made above may no longer obtain in full quantum gravity. The argument as it stands will hence not make any converts. Despite its appealing reliance on deeply entrenched physical principles, the argument thus falls short of proving that spacetime must be discrete (or, similarly, that gravity must be quantized) from the resources of trusted physical theories alone.
>
> (Wüthrich, 2005, p. 782)

In other words, the Heisenberg collapse argument may show that combining a classical notion of spacetime with quantum aspects is inconsistent, but this does not establish any specific positive conclusion for ontology. This is correct and also precisely what I take Arkani-Hamed's version of the Heisenberg collapse argument to state. Note, however, that Wüthrich takes this to induce the general question of

> how *any* argument drawing solely on accepted physical theories can possibly establish that gravity must be quantized. If a quantum theory of gravity would be part of the established corpus of theories, the proof would be easy. But alas, it is not! The failure of current physics to offer a straightforward and unique path to a quantum theory

of gravity strongly suggests that the formulation of such a theory will require new physics. In this case, however, one cannot accept an argument from the resources of *old* physics alone to the effect that gravity must be quantized.

(Wüthrich, 2005, p. 782)

To this I would counter that Wüthrich demands too much. After all, even a standard quantisation method is indeed inductively constructing a *new* theory, not deducing one (Linnemann, 2019, Chapter 5). Still, well-established physics is obviously important for this – at least as a methodological guide, but also as a general constraint: well-established physics needs to be (approximately) preserved in any future theory, which turns out to be a highly constraining condition; "we cannot just dream up anything" (Matsubara, 2017, p. 2).

To conclude, it is fair to say that a majority of physicists do not take semi-classical theory proposals seriously – although, strictly speaking, they may not have been ruled out yet.[26] Instead, the many theoretical issues that are considered to be related to finding a theory of quantum gravity, the plethora of thought experiments similar to the ones above, and further theoretical problems of semi-classical theories, typically convince physicists that a full-fledged theory of quantum gravity that involves some sort of "quantisation" of general relativity is needed.[27] What is more, recently table-top experiments have been proposed independently by Bose et al. (2017) and Marletto and Vedral (2017) to indirectly probe the quantum nature of gravitation based on measuring gravitationally induced entanglement of mesoscopic test masses in adjacent matter-wave interferometers (for a philosophical assessment see Huggett et al., 2023; see also Adlam, 2022). It is argued that "if the mutual gravitational interaction entangles the state of two masses, then the mediating gravitational field is necessarily quantum mechanical in nature" (Bose et al., 2017, p. 240401-2).[28]

Nevertheless, it is equally fair to say that there is no consensus at all on the *concrete* conclusions one is supposed to draw from all these arguments with respect to theory building, as is apparent in the variety of proposals for a theory of quantum gravity which all suffer – albeit not equally severely – from theoretical problems.

5.1.3   Brief Review of Theory Proposals

This section attempts to present a non-technical overview of some of the most important approaches to quantum gravity.[29] The aim is to briefly analyse and compare the respective methodologies and ontological commitments. A rather concise survey of such a diverse and intensely debated field cannot be exhaustive. In my presentation of different theory proposals,

I will therefore restrict myself to specific aspects that, on the one hand, make transparent and do justice to the rationality of the overall idea, methodology, and internal logic of each research programme, and, on the other hand, also underline whether and how the different approaches are theoretically related. I shall primarily seek to set out the central insights, paradigms or (guiding) principles that motivate the different research programmes, and indicate why they are considered the best choices to build a research programme. Obviously, this is closely tied to the question of what the problem of quantum gravity is considered to be in the first place. Furthermore, I shall briefly mention the central achievements and open questions of the approaches.

Note that, despite their diversity, all approaches claim to be "conservative". After all, every approach starts off somewhere in known territory, and tries to employ established knowledge from quantum physics or relativity theory (e.g., core principles like Lorentz invariance and methodologies like quantisation) to explore what lies beyond; in this rough sense all approaches are conservative indeed. For example, the semi-classical approach is conservative in the sense that it does not conceptually modify our best theories, but simply tries to find the correct law-like connection. The reason for this tendency to retain the known seems to be that

> [i]t is not easy to come up with anything that does not immediately come into conflict with what we know from quantum physics and relativity. So, although there are a number of different approaches, we cannot just dream up anything.
>
> (Matsubara, 2017, p. 2)

5.1.3.1 *Effective Quantum Field Theory of Gravity*

A natural way to approach the construction of a theory of quantum gravity is to write down the quantum field theory for a massless spin-2 particle, the so-called graviton (see Section 5.2 for more details). This uses the standard perturbative approach to quantum field theory, where the first-order terms receive higher-order corrections via loop effects. Perturbative quantum gravity employs covariant quantisation techniques which maintain full Lorentz invariance throughout the quantisation procedure; the same is true for path integral methods (Kiefer, 2007).

The key problem is that this straightforward quantisation of general relativity is not predictive at high energies – technically speaking, the theory suffers from ultraviolet divergences, i.e., the theory is not UV-complete (Crowther & Linnemann, 2019). Essentially, this is because at energies close to the Planck scale the full perturbation series (including infinitely many higher-order contributions) becomes relevant – which requires the experimental input of an infinite number of parameters.[30] The perturbation series can only be controlled by introducing a cut-off beyond which the theory is not applicable. Below the cut-off, only a finite number of parameters need to be determined experimentally and the theory remains well-defined and predictive – in principle, this can be done to any finite order in the expansion.[31]

The fact that perturbative quantum gravity is already worked out in full as an effective theory – which accommodates all presently known experimental data – is sometimes underappreciated, especially in philosophy or popular science presentations:

> A lot of portentous drivel has been written about the quantum theory of gravity, so I'd like to begin by making a fundamental observation about it that tends to be obfuscated. *There is a perfectly well-defined quantum theory of gravity that agrees accurately with all available experimental data.* ...
>
> ...
>
> ... All the classic consequences of general relativity, including the derivation of Newton's law as a first approximation, the advance of Mercury's perihelion[32], the decay of binary pulsar orbits due to gravitational radiation, and so forth, follow ... within a framework in which the principles of quantum mechanics are fully respected.
>
> (Wilczek, 2002, p. 10)

In this perspective, the problem is not to come up with a theory of quantum gravity in the first place, but rather that the theory we already have at our disposal is not UV-complete – and it is this which sets off the whole research programme of quantum gravity.[33]

Although many issues regarding non-renormalisability have been tamed by better understanding the physics of effective theories, many would not consider this satisfying. In particular, the cut-off – on which all physical observables then depend – is unexplained. The theory requires completion, for example, by introducing more degrees of freedom to cancel the unwanted divergences that would render the theory non-predictive without the cut-off machinery. In addition, Wald (1984, p. 384) complains that "the breakup of the metric into a background metric which is treated classically and a dynamical field ..., which is quantized, is unnatural from the viewpoint of classical general relativity". This objection expresses the standard lore that the quantum field theory of gravity merely quantises gravity, not spacetime, and that spacetime remains fundamental. In continuation to what I put forward in Chapter 4, I reject both Wald's objection and the standard lore (see Sections 5.2 and 5.3). Given the classical spin-2 formulation of general relativity, the "breakup" is *not* "unnatural".

More importantly, what Wald hypostatises as a "classical background" is, in fact, an artefact of a specifically geometrical interpretation of general relativity – in a dynamical reading, the "classical background" is merely a redundant way of talking about certain dynamical symmetry properties of the matter fields. This highlights that there *is* a sense in which the quantum field theory of gravity can contribute decisively to the debate on the non-fundamentality of spacetime, *although* it is not a full-fledged theory of quantum gravity, but only an effective theory for the low-energy (large-scale) regime. In contrast to other approaches to quantum gravity, the quantum-field-theoretic approach does not rely on speculations about what might replace small-scale spacetime, but challenges spacetime fundamentalism at the level of well-understood *large-scale* spacetime – again, *independent* of seeking a potential completion with regard to small-scale spacetime. This is what makes this approach so interesting and particularly fruitful from a philosophical perspective.

5.1.3.2 Asymptotic Safety

Asymptotic safety was originally proposed by Weinberg (1979) and takes its departure precisely from the effective theory's suffering from inconsistencies beyond the cut-off scale. In a sense it is the most straightforward attempt to solve the renormalisation problem and thereby complete the quantum-field-theoretic approach. The general idea is that there is some constraint on the infinite number of parameters that renders the theory predictive. Concretely, one assumes that the perturbation series is controlled at high energies such that the infinitely many parameters are not independent, but are fixed by a so-called *fixed point*.[34]

If one succeeded and showed that there is such a non-trivial UV fixed point to which general relativity is a perturbation, then general relativity would, in fact, be the classical limit of a *renormalisable* quantum field theory "and we were deceived by using perturbation theory in the wrong variables" (Shomer, 2007, p. 8).

However, there are independent reasons to believe that this is not the case. For example, asymptotic safety is standardly viewed to be in tension with results from black hole thermodynamics (Shomer, 2007). It is usually assumed that the Bekenstein-Hawking entropy of black holes in general relativity (which is proportional to the surface area of a black hole's event horizon) should correspond to a microscopic entropy of the black hole's quantum degrees of freedom. But according to quantum field theory, the microscopic entropy scales with volume, not area. Thus, quantum gravity cannot be a quantum field theory at high energies.[35] This contradicts the asymptotic safety scenario of quantum gravity. It has been argued, however, that this objection falls prey to semi-classical presuppositions

(Doboszewski & Linnemann, 2018); what is more, the claim that asymptotic safety is contradicted by entropy arguments has been explicitly countered in the asymptotic safety literature (e.g., Falls & Litim, 2014).

On the practical side, if one does find a UV fixed point for pure gravity in four dimensions, one still needs to add matter to the theory. This, however, will generally change the running of the coupling parameters and the fixed point for the pure gravity case will be missed. Accordingly, matter should better be taken into account right from the start, which makes the computation more complicated.

Ontologically, asymptotic safety seems to be rather similar to quantum field theory, at first. Most importantly, asymptotic safety does not render spacetime as such non-fundamental (Friederich, 2018). Beyond the Planck scale, however, asymptotic safety has new results to offer. In particular, asymptotic safety seems to imply that spacetime is two-dimensional with fractal-like dynamical properties (Friederich, 2018; Niedermaier & Reuter, 2006).

5.1.3.3 *String Theory*

Another approach that is often argued to be continuous to the quantum field theory programme (e.g., Dawid, 2013; Kiefer, 2007; Weinberg, 1999) is string theory.[36] String theory is without doubt *the* mainstream approach to full quantum gravity. But it is more than just a theory of quantum gravity. String theory offers a unified description of all fundamental interactions and matter.

Put crudely, the basic idea for solving the renormalisation problem of quantum field theory is to introduce additional underlying degrees of freedom, i.e., to replace the concept of fundamental point-like particles by higher dimensional objects – in particular, one-dimensional strings. The different fundamental particles can then be characterised as different excitations of the one-dimensional strings.

A central aspect of string theory is that its fundamental objects are defined ("live") in a higher-dimensional spacetime.[37] So, to obtain a four-dimensional spacetime, spurious dimensions need to be "compactified" ("curled-up"). For example, by curling up a two-dimensional plane to form a cylinder (i.e., by identifying two opposing sides), one dimension is compactified. By then forming a torus (by identifying the opposing ends of the cylinder), the other dimension is compactified. The so-called compactification radius determines the energy scale at which the compactified dimension is visible. Since we do not have any experimental hints at additional compactified dimensions, such radii need to be very small. Note that we might live in a compactified space ourselves, with very large compactification radii (Huggett, 2017).

Another key feature of string theory that has various interesting philosophical implications is that there are actually several variants of the theory which are related by certain duality transformations (see Rickles, 2011 for an overview). For instance, Huggett (2017) argues that due to one of those dualities the space in which the strings evolve (the so-called "target space") should *not* be identified with the observed, phenomenal space of relativity that we experience; phenomenal space then is a non-fundamental, higher-level phenomenon. The issue is, roughly speaking, that the T-dual theories "attribute radically different radii to space – larger than the observable universe, or far smaller than the Planck length" (Huggett, 2017, p. 81). Huggett argues that therefore the radius of target space is actually indeterminate. On the other hand, Huggett continues, phenomenal spacetime does have a determinate radius. Hence, phenomenal spacetime is not target space, but a higher-level structure.

Notably, there is also the conjectured gauge/gravity duality – or, anti-de Sitter/conformal field theory (AdS/CFT) correspondence. According to the AdS/CFT correspondence, a string theory in anti-de Sitter spacetime is dual to (read: empirically equivalent to) a conformal quantum field theory without gravity on the asymptotic boundary of the anti-de Sitter spacetime (e.g., de Haro, 2017; de Haro et al., 2016; Rickles, 2013; Vistarini, 2019).

Although string theory is significantly more worked out than asymptotic safety and the other approaches to follow below, it still has its problems. Essentially, it turns out to be very difficult to extract an empirically adequate model of our world from string theory (which is technically related to compactification). Notably, this is not the main criticism, though. Such a model is generally expected to be part of the solution space of string theory, the problem is that tons of other models are as well (the so-called string theory landscape).

5.1.3.4 *Loop Quantum Gravity*

The previous three approaches are conceptually closely related, and can be argued to share a quantum-field-theoretic paradigm. In particular, they focus on quantising gravity. Spacetime is not quantised in these theory proposals. This is different for the main competitor of string theory, loop quantum gravity. The general idea is that general-relativistic spacetime "emerges" from an underlying discrete spin network structure (see Figure 5.1). Loop quantum gravity essentially comes in two methodological variants that use either of the two standard quantisation methods: canonical loop quantum gravity and covariant loop quantum gravity. Let me restrict the presentation to the former (which is the traditional version).

Canonical loop quantum gravity attempts to quantise general relativity according to the so-called canonical quantisation method. This method

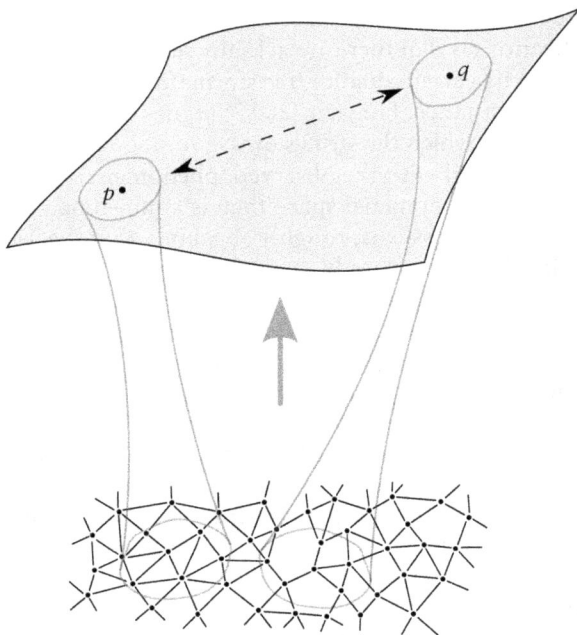

Figure 5.1 Emergence of general-relativistic spacetime from a spin network (own recreation of Christian Wüthrich, *A spacetime emerging from a spin network* in Wüthrich (2018, p. 322, Fig. 25.2); reproduced with permission of the Licensor through PLSclear). Note that fundamentally adjacent parts of the spin network (i.e., parts that are connected by an edge) may give rise to widely separated regions of spacetime (Huggett & Wüthrich, 2013).

starts from a Hamiltonian formulation of general relativity (as other canonical quantum gravity approaches do as well). A Hamiltonian system obeys certain differential equations (namely, Hamilton's equations) which relate generalised position and momentum variables (the so-called canonical variables) of all physical degrees of freedom to the energy of the system. Hamilton's equations then give the temporal evolution of the degrees of freedom of the system. Standardly, general relativity is not a Hamiltonian system, because there is a sense in which the Einstein field equations are not dynamical equations (that articulate well-posed Cauchy problems, i.e., initial value problems) but local constraint conditions on pairs of values of the metric field g and the energy-momentum tensor T (Wüthrich, 2012a). To obtain a dynamical formulation of general relativity, one essentially needs to split spacetime into spacelike hypersurfaces evolving over time. This so-called "foliation of spacetime" into three-dimensional spaces ordered by a one-dimensional time parameter then facilitates a Hamiltonian formulation. Accordingly, Lorentz covariance is not manifest in the quantisation procedure. The Hamiltonian formulation is accompanied by additional constraint equations, which the canonical variables need to obey, such that the theory is indeed equivalent to Einstein's theory of general relativity. The constraint equations basically ensure that the initial data cannot be chosen arbitrarily. When the theory is quantised,[38] the constraint equations are dragged along: only those quantum states which satisfy the quantum constraint equations are allowed. The Wheeler-DeWitt equation $\hat{H}(x)\psi = 0$ is such a constraint equation. Notably, $\hat{H}$ and ψ are *not* the familiar objects from quantum mechanics (Unruh & Wald, 1989).

Loop quantum gravity is standardly argued to suggest several significant ontological consequences (see Figure 5.1). It arguably suggests some kind of "emergence of spacetime" scenario where spacetime disappears at the fundamental level and only emerges from an underlying quantum structure (e.g., Huggett and Wüthrich, 2013). In particular, time is taken to "disappear".[39]

Technically, the disappearance of time can be traced to the solution space of the Wheeler-DeWitt equation. Essentially, the Wheeler-DeWitt equation renders the allowed quantum states time independent. Moreover, as $\hat{H}$ commutes with all observables, the dynamics is "frozen". Change appears to be pure gauge redundancy. In this sense, the theory predicts a fundamentally timeless world and does not allow for change – time and change disappear, which prompts speculations that a Parmenidean view receives support from canonical (loop) quantum gravity (e.g., Huggett et al., 2013; Wüthrich, 2012a).

Notably, the issue of the disappearance of time and change is present at the classical level as well. For classical Hamiltonian general relativity, the Hamiltonian H features in a constraint equation – the classical version of the Wheeler-DeWitt equation – which leads to the dynamical "freezing" of the system. Huggett et al. (2013) conclude that at least one version of the "problem of time" already arises at the classical level. Moreover, as the theories are supposed to be equivalent, this should not only be a feature of Hamiltonian general relativity, but of standard general relativity as well (see Earman, 2002; but cf. Healey, 2002, 2004; Maudlin, 2002; see also Huggett et al., 2013) – how could a mere change of representation bring about an interpretational issue that significant? In particular, the constraint equations, which result in the problem of time appearing, are there only to make the Hamiltonian formulation equivalent to Einstein's theory. Conversely, if, as Wüthrich (2012a) argues, general relativity does allow for some form of change (namely in terms of worldline evolution), there should be a sense in which Hamiltonian general relativity allows for change as well; for example, one might try to conceptualise change in

Hamiltonian general relativity with respect to the external foliation parameter. Note, however, that there is a deep-rooted tension between such proposals and the fact that general relativity is reparametrisation invariant, i.e., gauge invariant (Unruh & Wald, 1989). Time in general relativity is an arbitrary label assigned to some spacelike hypersurface. All physically meaningful quantities are independent of such labels – put formally, they must be diffeomorphism invariant. It is precisely these facts that are encoded in the constraint equations.

Proposals on how to solve the problem of time are due to Rovelli (2011), for example, who proposes a relational account of time's being non-fundamental, i.e., he proposes to interpret mechanics as a theory of relations between variables, rather than a theory of variables evolving in time.

5.1.3.5   Causal Set Theory

For the last proposal, recall the argument on the operational breakdown of spacetime (see Section 5.1.2). Causal set theory embraces in full the intriguingly simple conclusion that spacetime might be granular, not continuous, i.e., that there might be "atoms of spacetime" – arguably in analogy to the history of matter theories and the development of quantum theories of matter. Less figuratively speaking, causal set theory is motivated by results from general relativity regarding metric structure being almost completely determined by causal structure (see Malament, 1977; see also Dowker, 2013). Systematically, causal set theory stands somewhat apart from the other theory proposals presented; as causal set theory starts out from general relativity and postulates a discrete base structure (see Figure 5.2), it is closest to loop quantum gravity. Also, causal set theory is much less worked out than the other proposals; in particular, the biggest issue is that a quantum version has not been formulated yet. See Dowker (2013) for an accessible introduction to causal set theory and Wüthrich (2012b), Wüthrich and Callender (2017), and Lam and Wüthrich (2018) for a philosophical appraisal.

Ontologically, causal set theory is rather simple. It posits – notably, already classically[40] – a discrete and partially ordered fundamental structure: a set of elementary events that are related by a causal precedence relation (which is an antisymmetric, irreflexive, and therefore *asymmetric* relation between two events). Continuous (general-relativistic) spacetime is then supposed to emerge as the large-scale limit from the way the fundamental events combine due to the causal precedence relation.[41] Space and time, or spatiotemporal notions of any kind, do not feature at the fundamental level – with one exception: since the causal precedence relation is asymmetric, at least this key aspect of time remains fundamental. In this sense, time is explained by a fundamental causal relation, which is reminiscent of the tradition of causal theories of time (see Robb, 1914; Sklar, 1977). Space,

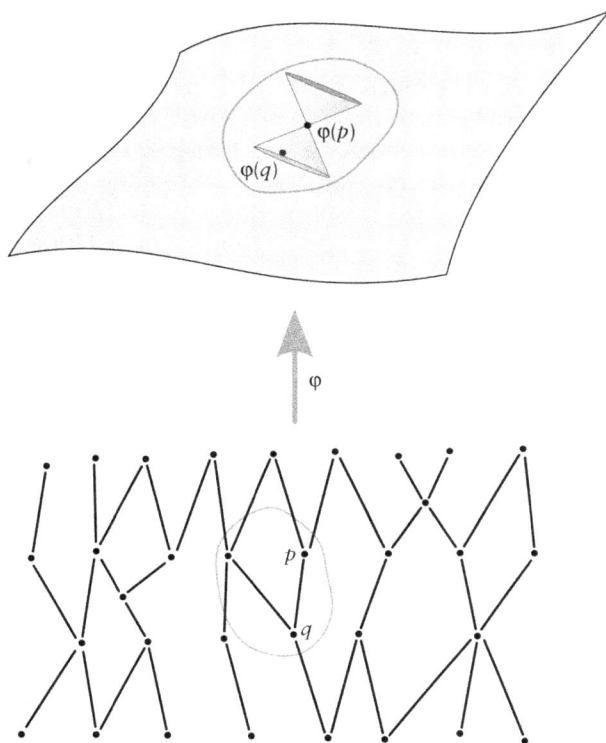

Figure 5.2 Emergence of general-relativistic spacetime from a causal set. General-relativistic spacetime with causal light cone structure is obtained from a causal set via a map φ (own recreation of Christian Wüthrich, *A causal set and an emergent spacetime related by a map* φ in Wüthrich (2018, p. 320, Fig. 25.1); reproduced with permission of the Licensor through PLSclear). Note that the fundamental causal structure needs to be preserved by φ; in particular, if an elementary event *q* causally precedes another elementary event *p*, then φ(*q*) needs to be in the past light cone of φ(*p*) (see Wüthrich, 2018).

however, has disappeared altogether (Lam & Wüthrich, 2018). Here is why. Suppose one picks one of the fundamental events as the "here-now". Then space is the maximal set of events that contains the "here-now" and all events that are *not* causally related to the "here-now". Since two events having no causal relation implies the two events having no relation at all, the fundamental space of causal set theory has no structure at all – no metrical structure, no conformal structure, no affine structure, no topological structure; the fundamental "space" of causal set theory does not even have dimensionality (Lam & Wüthrich, 2018).

A concern that is frequently voiced against theory proposals that replace spacetime by a fundamentally discrete structure is that discretisation compromises an important symmetry of physics, namely Lorentz invariance. Roughly speaking, the issue is that a fundamental discrete lattice structure picks out a preferred reference frame – thereby, violating Lorentz invariance. Indeed, some embrace this and investigate violations of local Lorentz invariance in quantum gravity. However, proponents of causal set theory claim that local Lorentz invariance can in fact be retained for a random distribution of the discrete set structure's points (Dowker et al., 2004).[42]

5.1.4 *Implications for the Fundamentality of Spacetime*

When evaluating these proposals, it becomes apparent that they do not have much in common. They suggest entirely different fundamental entities and very different ways of how to recover the spacetime of general relativity as a non-fundamental higher-level structure. In particular, the "different theories leave more or less of the standard structure of spacetime intact" (Huggett & Wüthrich, 2013, p. 277).[43]

On the one hand, there are proposals which straightforwardly reject the idea that spacetime or indeed any type of spatiotemporal structure is fundamental: proposals like loop quantum gravity and causal set theory do not seem to feature fundamental spatiotemporal notions, but posit an underlying discrete (quantum) structure from which spacetime and (almost) all its properties are supposed to arise. According to these theories, spatiotemporal structure *as such* is reduced to (or emergent from) underlying non-spatiotemporal structure.

On the other hand, theories like string theory challenge the fundamentality of spacetime to a lesser degree by proposing ways to obtain, for example, metrical structure from conformal structure (AdS/CFT correspondence) – conformal structure qualifies as less spatiotemporal than metric structure, but is still spatiotemporal; four-dimensional spacetime from ten or eleven dimensional spacetime – only the dimension of spacetime is challenged, not spatiotemporality as such; or phenomenal spacetime from target space – amongst others, the fundamental target space might lack certain features we usually conceive of as typical for spaces (e.g., a determinate radius), but is spatiotemporal otherwise.[44] String theory thereby retains more spatiotemporal structure than, for example, causal set theory and loop quantum gravity. Hence, string theory antagonises spacetime fundamentalism more moderately, and in a way that resembles past changes to our picture of spacetime – for instance, when moving from Newton's to Einstein's theory.[45]

What is more, my dissent has not (yet) changed the fact that the quantum-field-theoretic spin-2 approach is deemed fully spatiotemporal by most philosophers; namely, to only quantise gravity but retain fundamental Minkowski spacetime. The approach is usually not even discussed when it comes to the question of whether spacetime is fundamental in theories of quantum gravity (e.g., Huggett & Wüthrich, 2013; Matsubara, 2017).

In summary, the diversity of approaches prevents us drawing positive metaphysical conclusions of the form "quantum gravity suggests this and that type of fundamental structure". Without some criteria for how to delimit the theory space considered, *all* approaches to quantum gravity equally determine the metaphysics of spacetime – especially according to what has been dubbed *inductive metaphysics* (see Chapter 6). What all approaches agree on defines a *common core* and thereby constrains the metaphysical conclusions: the metaphysical status of spacetime is determined *negatively* by what is ruled out for not meeting the common core constraint (as opposed to being determined positively by positing a specific ontology that is, for example, "read off" from an approach or inferred otherwise).

More concretely, given that various approaches retain certain aspects of fundamental spatiotemporal structure (some maybe even full Minkowski spacetime), one cannot generally conclude that quantum gravity does away with substantival spacetime or spatiotemporality as such – in contrast to what is often claimed. Here, I agree with Esfeld (2021). The only metaphysical claim suggested unanimously by the various approaches seems to be that specifically *general-relativistic* spacetime is non-fundamental – presumably in all championed variants of that notion. Some *other* spacetime may very well be fundamental; or, in light of Jaksland and Salimkhani (2023): a different set of spatiotemporal aspects may very well be fundamental. Only speculative theory proposals, which are not worked out in full yet, seem to go beyond this.

Now, this conclusion may run against both general-relativistic and quantum-gravitational orthodoxy, but other than that, it seems that nothing profoundly new follows. Without further argumentation, for example, Minkowski spacetime is just as spatiotemporal as general-relativistic spacetime; we would merely settle for a different fundamental spacetime than previously thought. So, further argumentation is key.

One option is to aspire to obtain more substantial results by restricting the set of theories on which the metaphysical conclusions are based. For instance, one might dismiss the quantum field theory of gravity, because it is only an effective theory: it only applies to large scales which does not help us understand the small-scale aspects of quantum gravity. Or, relatedly, that a fundamental Minkowski metric still falls prey to Heisenberg-collapse-style arguments. In other words, one might argue that, ultimately, only UV-complete theories of quantum gravity delimit the metaphysics of spacetime. Dismissing quantum spin-2 theory would then leave string theory as posing the strictest constraint on non-fundamentality claims regarding spatiotemporality.

This procedure carries the danger of engineering metaphysical conclusions by excluding theories on the basis of contestable criteria. I therefore propose to not discard any (empirically) tenable approach from the start.[46] In fact, I propose to not exclude but rather focus on the quantum spin-2 approach. This is for two reasons: on the one hand, it is the *only* non-speculative, actually working theory of quantum gravity for the time being (albeit incomplete); on the other hand, it is the most restrictive approach with respect to the status of spatiotemporality. Hence, taking the quantum spin-2 approach into account makes any metaphysical inference more robust. Only afterwards, and on the basis of physical practise, shall I then propose a more refined and less engineered criterion to constrain metaphysical inferences.

Focusing on the quantum field theory of gravity, which, again, is only an effective theory for large scales, is justified especially if we consider spacetime to be a "package deal" as discussed above: when spacetime's fundamentality is called into question at some scale, it is called into question at all scales. If spacetime is non-fundamental at large scales, it is non-fundamental at small scales – independent of what might complete the effective theory in the small-scale regime.

So in the following, I shall not dwell further on questions that are already amply addressed in the literature on how the individual speculative theories of quantum gravity replace spacetime. Instead, I propose to build on my results from Chapter 4 and investigate more closely what the most spatiotemporal and least speculative theory, the quantum field theory of gravity, can contribute to the debate when interpreted dynamically. Recall that if a fixed-field formulation is feasible, metrical structure *as such* can be seen as arising from matter field dynamics. Pending further arguments in support of this dynamical view, the non-fundamentality of the g field is at least one tenable interpretation of general relativity. With respect to general relativity, potential further arguments are that the dynamical view is more coherent and has higher explanatory strength than the standard fundamentalist interpretation of general relativity. With respect to quantum gravity, the Heisenberg collapse argument provides further support for the dynamical perspective: in a quantum context, the interpretation of the Minkowski metric η as a mere codification of Lorentz symmetry is preferred since the Heisenberg collapse argument tells against *any* form of fundamental continuous metric structure. Based on these considerations, I shall advance a dynamical take on the quantum spin-2 theory of gravity below.

5.2   The Spin-2 Approach to Quantum Gravity

Given that the standard interpretation of the quantum spin-2 theory of gravity imposes the strictest constraint on the common core, showing what spatiotemporal structure is non-fundamental on this account can provide us with an upper bound on fundamental spatiotemporal structure. Specifically, I argue that metric structure is non-fundamental, making this upper bound less spatiotemporal than usually observed. This also serves to demonstrate that claims about the non-fundamentality of spacetime in quantum gravity do not rely solely on speculation about new physics.

With respect to the general rationale, the following is already familiar from Chapter 4 on the classical version of spin-2 theory; but it is the quantum version that constitutes the conceptual basis for this, and, or so I shall argue, more robustly determines the ontology.

5.2.1 *Preliminaries*

As mentioned above, the spin-2 theory of quantum gravity can in principle be obtained from a covariant quantisation of general relativity. It is usually presented as starting out from "splitting up" the metric field g of general relativity into a background metric η (or generally any solution $\tilde{g}$ of the Einstein field equations) and a linear perturbation, i.e., a second-rank tensor field h (e.g., Wald, 1984); as we have already seen in Chapter 4, more generic field-theoretic considerations – independent of presupposing general relativity – also lead to the same classical result. Then, the field h is quantised (as a non-abelian gauge theory). Notably, the fixed background η is standardly stressed to remain classical (Wald, 1984) – which I shall contest below.

So quantised, the theory agrees with the results of classical general relativity and all observational data in the low-energy regime (i.e., for long distances). But for high energies (small distances), perturbative quantum gravity is non-predictive and requires modifications: the covariant quantisation approach runs into the well-known problem of the theory's non-renormalisability.

In this section my aim is to present the conceptual basis behind the covariant quantisation approach following the presentation of Feynman et al. (1995), Hatfield (1995), and Duff (1975). Indeed, there is a conceptual framework that allows for building a theory of quantum gravity "from the bottom up" (Hatfield, 1995, p. xxxiii) without relying on general relativity and a quantisation prescription, and without directly running into the high-energy problems of local field theory (Boulware & Deser, 1975, p. 194), namely by using only the experimentally established phenomenological properties of gravity and

> the firmly established principles of special relativistic quantum particle (rather than field) theory, namely Lorentz invariance and the postulate that all forces are transmitted by the virtual exchange of particles.
>
> (Boulware & Deser, 1975, p. 194)

Recall that quantum particles generally have the following fundamental properties[47]: mass, charge, and spin; including particles that have zero mass, carry no charge, or have spin zero. In a special-relativistic quantum particle theory framework, the interactions between two particles are represented by the exchange of mediator particles – for the interactions of the Standard Model of particle physics these are the "gauge bosons". For example, the electromagnetic interaction between two electrons is understood as an exchange of a massless spin-1 particle, the photon (see Figure 5.3).

The mediator particle that is postulated to represent gravity is the *graviton*, a massless spin-2 particle. The general idea of the "bottom up" approach is to first identify the fundamental properties of the graviton from the established phenomenological facts about gravity alone, i.e., independent of a fully worked out theory. Second, the graviton is then studied in the generic special-relativistic quantum particle framework with respect to potential additional, derivative properties. Only then is it that one realises that the full dynamics of this theory has general relativity as its classical limit.[48]

In the more familiar case of electromagnetism, the procedure starts as follows. First, it is postulated that the electromagnetic interaction of, say, two electrons is represented by the exchange of a mediator particle, namely the photon. Second, the fundamental properties of this mediator particle are determined by observing basic facts about the phenomenology of electromagnetism: (a) electromagnetism is an interaction of infinite range, which gives that the photon is massless; (b) electromagnetism does not only produce scattering processes, but is a static force as well, which gives that the photon has *integer* spin[49]; and (c) like electromagnetic charges repel each other, which gives that the photon has *odd* integer spin. Notably, this does not yet fix whether the photon has spin one, three, or higher. I come back to this below.

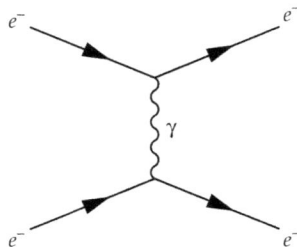

Figure 5.3 Feynman diagram (t-channel) for Møller scattering of two electrons, $e^-e^- \to e^-e^-$. The electromagnetic interaction between the two electrons e^- (straight lines) is mediated by the exchange of a photon γ (wavy line).

Gravity, just like electromagnetism, is a long-range force that obeys an inverse square law of infinite range. Therefore, the graviton is massless. Second, just like the photon, the graviton needs to have *integer* spin to be able to produce a static gravitational force (Hatfield, 1995, p. xxxiv). Finally,[50] unlike electromagnetism, gravity is attractive only, which further determines the graviton's spin to *even* integer spins, i.e., spin zero, two, four, or higher.[51] The possibility of a spin-0 graviton can be excluded by the empirical result that photons are affected by gravity (e.g., bending of light in gravitational fields).[52] Further restrictions that exclude higher spin gravitons are discussed in the next section.

With regard to the general rationale, the different quantum derivations then proceed in essentially the same manner as in the classical case: the consistency of some general set-up for an *interacting* graviton requires specific coupling properties. Similar to the classical case, universal coupling of the graviton to *all* particles that have energy and momentum is established as the result of postulating consistent Lorentz-invariant graviton interactions with *some* matter. In particular, the essence of Weinberg's *soft graviton argument* – where "soft" refers to the graviton's being low-energetic – is that

> we cannot construct a Lorentz-invariant theory of massless particles with spin $s \geq 1$ without, at the same time, building some sort of gauge invariance into the theory.
>
> (Duff, 1975, p. 83)[53]

5.2.2 Weinberg's Soft Graviton Argument

Weinberg's argument is essentially concerned with deriving consistency requirements for physical processes from demanding Lorentz invariance; here physical processes are generic scattering processes of relativistic quantum particles. There are several ways to articulate the argument, or put forward similar arguments.[54] For his version, Weinberg considers a generic Lorentz-invariant scattering process $\alpha \to \beta$, where α denotes the incoming and β the outgoing relativistic quantum particles i of momenta p_i^μ (see Figure 5.4a).

Figure 5.4a shows the standard way of conceptualising such a scattering process: roughly speaking, a bunch of particles come in from infinity, interact in some way (which is represented by the grey circle), and then a bunch of particles (the same particles or others) go out to infinity again. For any such process we can write down a so-called *scattering amplitude* $\mathcal{M}_{\alpha\beta}(p_1, \ldots, p_n)$, which is a complex number that represents the probability amplitude of the process.[55] Notably, no specification of $\mathcal{M}_{\alpha\beta}(p_1, \ldots, p_n)$ is needed for Weinberg's argument – hence the completely generic incoming

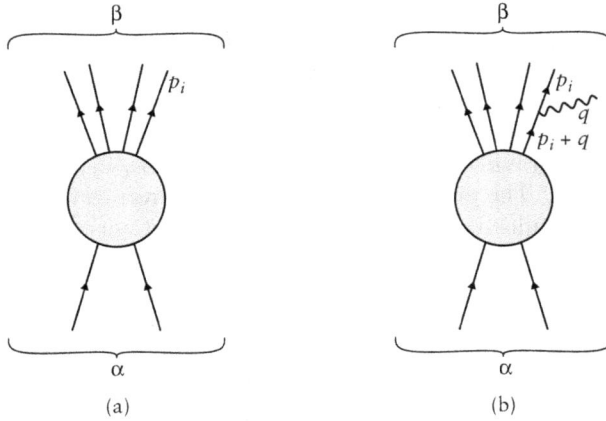

Figure 5.4 Weinberg's soft graviton argument. (a) Arbitrary scattering process $\alpha \rightarrow \beta$ (Weinberg, 1995, p. 536) and (b) dominant diagram for additional emission of soft photons or gravitons from an outgoing particle in an arbitrary scattering process (Weinberg, 1995, p. 536).

and outgoing states, and the grey circle as representation of the details of the interaction process. Weinberg's argument is entirely independent of which incoming and outgoing particles are considered and how they interact. In particular, the argument does not involve the presupposition of a Lagrangian. One only assumes that $\mathcal{M}_{\alpha\beta}(p_1,\ldots,p_n)$ does not vanish.

Without loss of generality, Weinberg then considers a slightly different process that additionally involves the emission of a massless quantum particle of arbitrary spin and with momentum q from one of the incoming or outgoing states (see Figure 5.4b). In particular, Weinberg considers the limit of $q \rightarrow 0$, i.e., the emission of a *soft* massless particle. For emphasis, "soft" means that the particle is supposed to have vanishing momentum, i.e., $q \rightarrow 0$. In this limit, the emitting particle i retains its momentum p_i. So, besides the emission of q, the process is the same, with the same incoming an outgoing momenta and spins (Nicolis, 2011, p. 19). Accordingly, the corresponding amplitude $\mathcal{M}_{\alpha\beta}(p_1,\ldots,p_n,q)$ is obtained from multiplying the original amplitude $\mathcal{M}_{\alpha\beta}(p_1,\ldots,p_n)$ by a non-trivial "emission factor" $\mathcal{E}$. The subsequent task is to analyse if and how the emission process and the corresponding interaction is constrained.

For clarification, here is why it is the soft limit of q that is interesting. When writing down the amplitude for Figure 5.4b, the propagator $p_i + q$ generally yields the following contribution (Arkani-Hamed, 2010a):

$$\frac{1}{(p_i + q)^2 - M_i^2},\tag{5.2}$$

where M_i^2 is the mass of particle p_i. In the $q \to 0$ limit, expression (5.2) becomes large, which is what we are interested in when studying the long-range behaviour. Note that q^2 can be neglected for $q \to 0$ such that expression (5.2) becomes[56]

$$\frac{1}{2\,p_i \cdot q},\qquad(5.3)$$

which diverges in the limit of $q \to 0$.[57] Hence, expression (5.3) will feature in the dominant leading-order contribution to this process (Arkani-Hamed, 2010a),[58] which determines the long-range behaviour.

Let me now analyse the properties and constraints such an emission process poses for photon interactions, specifically. Suppose $\mathcal{M}_{\alpha\beta}(p_1,\ldots,p_n)$ is the amplitude of some arbitrary scattering process, as in Figure 5.4a. To mathematically account for the additional photon emission, as in Figure 5.4b, the original amplitude needs to be multiplied by a non-trivial factor $\mathcal{E}_{\text{Photon}}$. Because we want to model long-range phenomena, we only consider the leading-order contribution that survives "at infinity", that is in the long-range limit. Generally, this factor will depend on expression (5.3), the polarisation vector of the photon $\varepsilon_\mu(q)$, all the momenta of the other particles p_i,[59] and generally some constant e_i, which can later be identified with the respective charge.[60] Accordingly, we obtain the following amplitude for the modified process of Figure 5.4b (Arkani-Hamed, 2010a):

$$\mathcal{M}_{\alpha\beta}(p_1,\ldots,p_n,q)\big|_{q\to 0} = \mathcal{M}_{\alpha\beta}(p_1,\ldots,p_n) \times \sum_i e_i \frac{p_i^\mu}{2p_i \cdot q} \varepsilon_\mu(q)\qquad(5.4)$$

The additional factor for the photon emission

$$\mathcal{E}_{\text{Photon}} = \sum_i e_i \frac{p_i^\mu}{2p_i \cdot q} \varepsilon_\mu(q)\qquad(5.5)$$

is constrained further: the additional emission of a soft photon should not spoil Lorentz invariance. Thus, we demand that Lorentz invariance is preserved.[61] As a result, Eq. (5.4) is required to become zero when replacing $\varepsilon_\mu(q)$ by q_μ (because shifting $\varepsilon_\mu(p)$ by anything proportional to p must give the same expression; see endnote 61 and Appendix A.2 for more details). Since the original amplitude $\mathcal{M}_{\alpha\beta}(p_1,\ldots,p_n)$ is assumed to be non-zero (the original process is not forbidden), the emission factor itself has to vanish:

$$\mathcal{E}_{\text{Photon}}\big|_{\varepsilon_\mu(q)\to q^\mu} = \sum_i e_i \frac{q^\mu p_{i,\mu}}{2p_i \cdot q} = \sum_i e_i \frac{1}{2} = 0.\qquad(5.6)$$

Accordingly, we obtain that the sum over all charges e_i needs to be zero,

$$\sum_i e_i = 0, \qquad\qquad (5.7)$$

which means that the process is forced to obey *charge conservation*. So, interactions with soft massless spin-1 particles always conserve the respective charge; for the photon case, this is the electromagnetic charge, of course. In fact, one could go on and derive Maxwell's equations by using perturbation theory (see Weinberg, 1965a). But let me stop here and turn to the next and, for my purposes, more interesting case: the emission of a massless spin-2 particle, commonly referred to as the graviton.

In the graviton case, the argumentation is essentially the same. Just like before, we want to study the leading-order contribution to the process in Figure 5.4b. We only need to replace the photon emission factor $\mathcal{E}_{\text{Photon}}$ by the graviton emission factor $\mathcal{E}_{\text{Graviton}}$ as follows (Arkani-Hamed, 2010a):

$$\mathcal{M}_{\alpha\beta}(p_1,\ldots,p_n,q)\Big|_{q\to 0} = \mathcal{M}_{\alpha\beta}(p_1,\ldots,p_n)\times \sum_i \kappa_i \frac{p_i^\mu p_i^\nu}{2p_i\cdot q}\varepsilon_{\mu\nu}(q). \qquad (5.8)$$

Here, $\varepsilon_{\mu\nu}(q)$ is the polarisation tensor of the graviton, and κ_i are the coupling constants for the particles with momenta p_i emitting a soft graviton (Weinberg, 1965a, p. B989). Again, the graviton emission factor in Eq. (5.8) is the leading-order contribution to the emission (as the long-range behaviour is of interest). Now, if we demand Lorentz invariance[62] and assume that $\mathcal{M}_{\alpha\beta}(p_1,\ldots,p_n)\neq 0$, we arrive at:

$$\sum_i \kappa_i\, p_i^\nu = 0. \qquad\qquad (5.9)$$

How to interpret this? According to Eq. (5.9) the sum over all momenta p_i weighted with the coupling constants κ_i is conserved in all possible scattering processes. However, (unweighted) momentum conservation, i.e., $\sum_i p_i = 0$, already holds in all scattering processes. Both conservation of momentum and Eq. (5.9) can only hold simultaneously if either the scattering between the particles of momentum p_i is trivial (the particles do not interact, i.e., the individual momenta p_i^μ do not change at all), or the coupling constants κ_i are *the same for all particles regardless of their properties*, i.e., $\kappa_i = \kappa$.[63]

The soft graviton argument shows that Lorentz invariance constrains massless spin-2 particles to *couple universally* to other particles.[64] Notably, no restrictions apply as to which particles are considered. Hence, the established universal coupling property of the graviton implies *graviton self-coupling*.

The derivations of the spin-2's universal coupling by Weinberg or Boulware and Deser have several other important features. For instance, they imply that the massless spin-2 particle is unique: if the coupling constant is universal for all particle species, "another" massless spin-2 field would have the exact same properties. Two universal interactions are indistinguishable. It also follows that gravitation cannot be a higher spin effect: there are no higher spin (tensor) contributions to the long-range behaviour of gravity,[65] because this would impose a further (weighted) constraint on the summation of the momenta, which could only be satisfied if the corresponding couplings vanished universally. Moreover, there are no conserved symmetric tensors of rank $r \geq 3$ that could act as a source – except for the total derivative of an asymmetric tensor (see Weinberg, 1965a; 1965b; 1995). However, these so-called "Pauli-type" tensors (after Pauli, 1941), which can appear in non-minimal coupling terms, only contribute to the high-energy behaviour, not to the infrared limit (Boulware & Deser, 1975, p. 194). Accordingly, particles of spin $j \geq 3$ can exist, but they cannot mediate inverse square law forces or generally have couplings that are present at low-energies. Another way to put this is that all these modifications will behave as high-energy corrections to the general-relativistic limit.

What is more, analogous to the approach presented in Chapter 4, one can derive the Einstein field equations as the classical limit of the theory (see Weinberg, 1965a). A few qualifications do apply, though. First, we have so far neglected a cosmological term. It can be included by means of a specific matter-like contribution to the energy-momentum tensor (Boulware & Deser, 1975, p. 197). Second, one might consider it a shortcoming of Weinberg's argument that it depends on a particular class of backgrounds. On the one hand, the geometry needs to be flat and Minkowskian, so that Lorentz invariance can be employed straightforwardly. On the other hand, the topology needs to allow for the construction of an S-matrix, which excludes topologies without an asymptotic regime (Davies & Fang, 1982, p. 472). However, as Davies and Fang show, Weinberg's derivation can be generalised by directly working with the action. Again, the starting point is the free field structure which "imposes a remarkably restrictive set of conditions on the structure of the higher-order terms" (Davies & Fang, 1982, p. 473). To lowest order, varying the full Lagrangian ($\delta L / \delta h_{\mu\nu} = 0$) yields the (free) Fierz-Pauli equation which, via the familiar divergence identity (see Section 4.1.2), forces the higher orders to be divergenceless as well, order by order (Davies & Fang, 1982). Third, all mentioned quantum derivations are manifestly perturbative. As such, they rely on certain analyticity assumptions (Boulware & Deser, 1975, p. 194). As mentioned before, this will typically restrict the solution space to analytic solutions. Other classes of solutions are, *prima facie*, not recoverable in a mathematically rigorous way. So, within such derivations one would need to have some independent argument for why the recovered solution space may, nevertheless, be extrapolated to the full solution space of general relativity. Fourth, similarly to the classical spin-2 approach, presumably only globally hyperbolic solutions can be accommodated. But as discussed in Section 4.1.4, having a restricted solution space is generally not considered to render a theory physically uninteresting *per se* – quite the opposite, if the restrictions only exclude solutions that are considered unphysical anyway (which would need to be investigated).

5.2.3   *The Equivalence Principle*[66]

But instead of reiterating how to obtain the Einstein field equations in (quantum) spin-2 theory, let me study in detail what said "universal coupling" is about. In the physics literature, the derivation of the spin-2's universal coupling property is usually interpreted as deriving the (strong) equivalence principle of general relativity. This is praised as another major result of spin-2 theory, alongside deriving Einstein's field equations, because the equivalence principle captures the essence of gravity and is in many ways the key to interpreting general relativity. In what follows, I examine whether the praise is justified.

Broadly speaking, the principle of equivalence theoretically records empirical facts about gravity's peculiar influence on physical bodies,[67] which have been known essentially since Galileo and can be expressed, for example, in terms of Einstein's elevator thought experiment (see Section 3.3). More specifically, "the" equivalence principle of general relativity comes in the following traditional versions, the *weak* (WEP-GR), the *Einstein* (EEP-GR), and the *strong* (SEP-GR) equivalence principle of general relativity[68]:

WEP-GR   The free-fall trajectories of test bodies are independent of the test bodies' composition.

EEP-GR   The non-gravitational laws of physics are the same in all local freely falling frames.

SEP-GR   Locally, all laws of physics reduce to the laws of special relativity.

For the sake of simplicity, I have focused here on the key aspects and subsumed additional qualifications into notions like "test body" and "composition", to which I turn below. Arguably, none of these versions is "traditional" in any historically accurate sense. In particular, none of them is Einstein's version (see Lehmkuhl, 2021). They are "traditional" in the sense that the division into a weak, a strong, and an intermediate version prevails in the literature. And in terms of content, while these three definitions do not capture all the subtleties of the debate, they do make some crucial distinctions (especially with regard to the aims of the following discussion). So, I urge the interested reader to keep that in mind and to consult the works cited as well; the experts I ask to bear with my slightly idiosyncratic presentation. In the following, I shall mostly be concerned with WEP-GR and SEP-GR.

According to Will (2001), WEP-GR

> states that if an uncharged test body is placed at an initial event in spacetime and given an initial velocity there, then its subsequent trajectory will be independent of its internal structure and composition. By 'uncharged test body' we mean an electrically neutral body that has negligible self-gravitational energy (as estimated using Newtonian theory) and that is small enough in size so that its coupling to inhomogeneities in external fields can be ignored.
>
> (Will, 2001, p. 16)

According to WEP-GR, which is enshrined in the geodesic principle of general relativity (Di Casola et al., 2015),[69] the free-fall behaviour of a test body is independent of its properties including its mass: all test bodies in a gravitational field fall at the same rate. This empirical essence of the weak equivalence principle is already familiar from the experimental results of Galileo with respect to Newtonian mechanics.[70] Notably, WEP-GR applies not only to point particles, but also to (sufficiently small) composite objects. However, the notion of a "test body" requires that the body is "passive" in the sense that it has no (relevant) back-reaction on its environment (e.g., Di Casola et al., 2015; Will, 2001). Thus, the notion of a test body involves idealisations (e.g., Di Casola et al., 2015). Its passivity implies that the test body cannot be charged (only its constituents can) and does not rotate or have spin. Then, all properties of the test body are fixed by fixing what constitutes it. After restricting the class of properties accordingly, one may then replace "independent of its composition" by "independent of its properties".

It is key to notice that the content of WEP-GR draws attention to two senses in which gravitation is *universal*: (1) *all* test bodies are affected and (2) they are so *in one and the same way*. As a result, the weak equivalence principle suggests the geometrisation perspective. Geometrisation explains why the response to a gravitational field is universal:

> Universality of free-fall indicates that the worldlines of test particles in a gravitational field do not depend on the particle properties, but only on their gravitational environment. Therefore, a gravitational field is associated with a set of preferred lines in spacetime, which can be considered the auto-parallel lines of some connection Γ (i.e., those

lines whose tangent vector, parallel-transported along the line itself, remains tangent). When there is no gravitational field, the preferred worldlines reduce to the straight-line inertial motions of Newtonian mechanics and special relativity, which correspond to a fixed, flat connection Γ_0. Conversely, in the presence of a gravitational field, the connection is, in general, dynamical and curved, as implied by elementary considerations about tidal forces.

(Di Casola et al., 2015, p. 42)

However, as also Di Casola et al. (2015) highlight, this identification of gravitation with spacetime geometry is based on a contestable philosophical presupposition, namely

a view in which all universal features are ascribed to geometry. According to this philosophy, the universal behavior of freely falling test particles is better explained by spacetime geometry than by a physical field (gravity) that couples to all bodies in the same way.

(Di Casola et al., 2015, pp. 42–43)

SEP-GR gives a rather different perspective on the physical content of "the" equivalence principle. It is about constraining the laws of physics (including gravity): the laws of physics are constrained to be locally special-relativistic, so that the local dynamical symmetries match the local spacetime symmetries[71]; here, "local" may be cashed out in a "pointy" or a "neighbourhood" version (see Read et al., 2018).

In terms of test experiments, this means that SEP-GR states that the results of local gravitational test experiments are independent of a potential gravitational background:

the physical processes considered here must be regarded as test experiments over a background that is not significantly affected by them. Thus, the kind of gravitational experiments to which the SEP [SEP-GR] applies can be, for example, the mutual attraction between two sufficiently light bodies, or the detection of a weak gravitational wave. The principle then says that, even if such experiments are performed within some background gravitational field, it is always possible to find observers who will (locally) record the same results, when performing the same experiments in the absence of such a background field. This is by no means obvious or automatic, in a nonlinear theory of gravity such as general relativity, where distant masses could have a nontrivial effect on local gravitational processes.

(Di Casola et al., 2015, p. 41)

Put differently, SEP-GR implies that any gravitational field can be "transformed away" locally such that the laws of special relativity hold (i.e., local experiments will confirm the laws of special relativity). This aspect of being able to locally "transform away" gravity and discover that spacetime is locally Minkowskian is intimately connected to the standard geometrical view that gravitation *is* curvature of spacetime. It brings with it the intriguing picture of the generally curved g field being flat when zooming in on small enough regions (or a point). However, this picture can be misleading: the curvature tensor may still be non-vanishing in this neighbourhood (or at this point) – g is *not* truly flat locally (which is empirically accessible, of course, by performing several spatiotemporally distributed local experiments). Therefore, it is better to state the physical content of SEP-GR as follows: for local physical experiments (including gravitational experiments), "the gravitational background is ... irrelevant – the outcome is indistinguishable from that of an identical experiment performed in Minkowski spacetime" (Di Casola et al., 2015, p. 44). The geometrically motivated idea that g locally "looks like" Minkowski spacetime is retained. Note that SEP-GR does not assume any restrictions regarding the types of experiments. It includes, for example, "processes involving the mechanics of particles and continua, thermodynamics, electromagnetism", and "the mutual attraction between two sufficiently light bodies" (Di Casola et al., 2015, p. 41). Accordingly, SEP-GR appears, *prima facie*, as a far more general statement than WEP-GR, which is restricted to experiments that probe the mechanical motion of a test body. In contrast to SEP-GR, EEP-GR applies only to the non-gravitational laws of physics and does not specify that these laws are special-relativistic. But there are also versions that do specify the latter (Di Casola et al., 2015, p. 41). In the literature, SEP-GR sometimes also appears as the "very strong equivalence principle" (Ciufolini & Wheeler, 1995, p. 40) to stress that it includes all laws.

Now, how are WEP-GR and SEP-GR related? Will explicitly defines EEP-GR (Will, 2001, p. 16) and SEP-GR (Will, 2001, p. 76) as "WEP-GR plus further conditions". I do not adopt this recursive view, since it presupposes what ought to be established by argumentation: the logical relation between the different versions of the equivalence principle. Still, the labels "strong" and "weak" and the apparently larger scope of SEP-GR do suggest that the (logical) relation between the two principles is obvious. SEP-GR is usually regarded as stronger than WEP-GR, in the sense that: (a) SEP-GR logically implies WEP-GR (and not *vice versa*), and (b) SEP-GR is more restrictive than WEP-GR.[72] But making such claims precise and robust is not trivial. This is because WEP-GR and SEP-GR rely on very different concepts that have yet to be articulated. WEP-GR talks about the properties of a test body and its free-fall motion, whereas SEP-GR talks about a local constraint condition on the laws of physics. Therefore, a blunt appeal to the apparent degrees of generality – WEP-GR seems rather specific, SEP-GR rather general – is insufficient. The difficult task is to show explicitly if one implies the other or not.

First, here is an argument why WEP-GR can be considered to *not* imply SEP-GR. Recall that WEP-GR states that the free-fall behaviour of a test body, which generally is a composite object, is independent of its composition. Consider a body that qualifies as a test body in the sense of WEP-GR – for example, a generic stone that is sufficiently small and sufficiently light not to back-react on its environment. WEP-GR states that the free-fall behaviour of the stone is independent of its composition. Arguably, the stone consists of fundamental particles, like electrons and quarks, and their interactions (most importantly, the electromagnetic and strong interaction, but in principle all interactions participate). In other words, the stone's composition is determined by the fundamental laws of physics. With this in mind, WEP-GR states that the free-fall behaviour is independent of any specification of the fundamental laws of physics. The free-fall behaviour is the same regardless of which laws govern the stone's constituents – in particular, regardless of whether these laws are special-relativistic or, say, Newtonian. SEP-GR, on the other hand, additionally requires that the fundamental laws are special-relativistic.[73] Hence, WEP-GR *does not imply* SEP-GR. The locality condition in SEP-GR does not seem to pose a further problem, since the restriction to test bodies already implies a notion of locality.

This argument seems convincing, but consider the following counter argument. The fact that the composition of the stone is determined by the fundamental laws of physics, so the argument goes, should not be read as "the free-fall behaviour is completely independent of which fundamental laws apply", but as "the free-fall behaviour *constrains the fundamental laws*": *the laws of physics must be so conditioned that they allow for the stone's composition not affecting its free-fall motion.* Why should we expect, so the reasoning goes, that WEP-GR holds for *all* types of fundamental laws? In other words, why should we expect that the Einstein field equations hold for any pair of independently chosen gs and Ts? Recall that the Einstein field equations are essentially local constraint conditions on pairs of values of the metric field g and the energy-momentum tensor T (Wüthrich, 2012a). After all, there is no empirical evidence that the internal structure of physical test bodies is determined by anything else than special-relativistic laws. We have no empirical evidence that WEP-GR holds for, say, fundamental Newtonian internal forces (let alone entirely different kinds of fundamental laws).

Such reasoning is reminiscent of Schiff's (1960) conjecture (see also Di Casola et al., 2015), "which states that any complete and self-consistent theory of gravity that satisfies WEP [the weak equivalence principle] necessarily satisfies EEP [Einstein's equivalence principle]" (Will, 2001, p. 12): "It is not obvious that, if such forces [that hold together the test body; my remark] were sensitive to the presence of a gravitational field" – which would mean that EEP-GR and SEP-GR would be false – "a universal behavior for 'test particles' [i.e., WEP-GR; my remark] could emerge" (Di Casola et al., 2015, p. 42). This open issue underlines that the logical relation between the two principles is non-trivial and depends on how terms like "composition" are specified. *If* something like Schiff's conjecture is true, then WEP-GR might actually imply SEP-GR.

It certainly seems to be a valid point that *some* conditions need to be fulfilled in order for WEP-GR to hold: suppose the most general version of WEP-GR were true, i.e., suppose free-fall motion was indeed entirely independent of any specification of the (laws of) composition, then there could be laws of composition that are sensitive to gravity and ones that are not. So the laws of composition would generally respond differently to the presence of a gravitational field. This would violate WEP-GR, since a test body that is composed according to the former class of laws would behave differently from a test body that is composed according to the latter class of laws. However, it seems unclear why only special-relativistic laws should allow for universal free-fall behaviour (which is what "WEP-GR implies SEP-GR" would mean). For instance, why should it not be a sufficient condition that all laws of composition (or, indeed, all fundamental laws) have the *same* symmetry properties (regardless of whether they are Lorentzian or, say, Galilean)?

Indeed, the laws must have the same symmetry properties, i.e., be invariant under the same transformations, because otherwise gravity would affect test bodies that are composed according to different laws differently – *contra* WEP-GR. This can be seen by noting one of the key insights of the debate on the dynamical approach: matter fields of different dynamical symmetries survey different spacetime metrics (see Section 3.5). Hence, the trajectories of "Lorentzian" stones are Lorentzian geodesics, whereas "Galilean" stones will traverse Galilean geodesics – in violation of WEP-GR. As Read points out: "[t]o move from ... results regarding the geodesic motion of small bodies, to the behaviour of matter fields *tout court*, is in effect to demand that the local symmetry properties of all matter fields be derivable from such geodesic motions; that is, it is, in effect, to demand a proof of a result akin to *Schiff's conjecture*" (Read, 2020, p. 198). If my above argument is correct, we can at least derive from the universality of geodesic motion of test bodies (i.e., WEP-GR) that the local symmetry properties of all matter fields must be the *same* – regardless of what type they are (so SEP-GR does not immediately follow).

Accordingly, this reasoning establishes that WEP-GR implies EEP-GR (i.e., Schiff's conjecture), which states precisely that all non-gravitational

laws are invariant under the same transformations. But it does not establish SEP-GR. Arguably, the restriction to *non-gravitational* laws can be
dropped, since gravitational interactions are in principle (albeit negligibly)
part of compositional laws. Then, Schiff's original conjecture is further
generalised – but still not to SEP-GR (which specifies the fundamental laws
as special-relativistic).

To summarise, if the above argumentation is correct, WEP-GR implies
more than usually acknowledged, namely that *all the fundamental matter
field dynamics are invariant under the same symmetry transformations*, but
it does *not* imply SEP-GR, since WEP-GR does not provide the resources to
additionally specify the type of invariance to Lorentz invariance. Presumably, it is this additional assumption of SEP-GR which singles out general
relativity.

Now, what about the reverse direction? Why should locally special-relativistic laws imply universal free-fall behaviour, as is the standard view?
Here is an argument that is independent of the potentially controversial
issues above. Suppose SEP-GR holds. Then local experiments will always
give the same results as experiments performed in Minkowski spacetime;
in particular, local mechanical experiments will always give the same results as experiments performed in Minkowski spacetime. In Minkowski
spacetime free motion of test bodies is universal (presumably, because
the underlying dynamics of their constituents is universally Lorentz-invariant). "Since the behavior of free particles is universal, it follows that
freely-falling particles behave in a universal way" (Di Casola et al., 2015,
p. 42). So if SEP-GR holds, then also WEP-GR holds.

To summarise, let me stress again that there are various versions of what
I have called WEP-GR and SEP-GR. They express the respective physical content slightly differently and their logical relations are not entirely
clear. Still, on a general level, I propose to say that "WEP-like" principles
refer to the fact that a body's response to gravitation is independent of
its properties or composition, and "SEP-like" principles refer to the fact
that the laws of physics are locally special-relativistic. Another way to put
this is that WEP-like principles "are … statements about properties and the
behavior of particular physical systems" (Di Casola et al., 2015, p. 43),
whereas SEP-like principles "can be reformulated as general 'impossibility
principles'", which "forbid the detection of a gravitational field by means
of local experiments". Since "body", "response", "properties", "composition", "locality", and "laws" can be specified differently,[74] one obtains
several versions of WEP-like and SEP-like principles of different strength.
Both WEP-GR and SEP-GR can be linked to a geometrical perspective on
gravitation, but it is SEP-GR which singles out general relativity (Di Casola
et al., 2015, p. 43). In particular, both principles rely on *geometrically or
spatiotemporally loaded concepts* like "trajectories" and "locality".

When turning to quantum field theory, or non-geometrical theories in general, which rely on concepts like "interaction" or "coupling", we would expect the geometrical coating to come off. For comparing the traditional principles in general relativity to similar principles in other theoretical frameworks, it is therefore useful to reformulate the traditional equivalence principles in (potentially more fundamental) non-geometrical terms – such as Lorentz invariance, universal coupling to g, and minimal coupling.[75]

To capture the general physical content of the equivalence principles independently of geometrical terms we usually allude to gravity's universality: *all* objects are subject to gravity in the *same* way. Recall that it is this property of gravitation that suggested geometrisation in the first place (in terms of spacetime being the reduction base in a reductive explanation). The fact that gravitational effects are apparently independent of the objects' properties supports the conclusion that gravitation arises from something else, namely spacetime:

> The property that all non-gravitational fields should couple in the same manner to a single gravitational field – universal coupling – allows one discuss the metric g as a property of spacetime itself rather than as a field over spacetime. This is because its properties may be measured and studied using a variety of different experimental devices, composed of different non-gravitational fields and particles, and because of universal coupling, the results will be independent of the device.
>
> (Will, 2001, p. 61)

But the universality of gravitation also resonates well with the framework of non-geometrical theories like quantum field theory, where we can account for gravity's universality more directly in terms of universal coupling (to the spin-2 graviton):

> In the usual deductive formulation of the general theory of relativity, the *a priori* identification of the metric of space with the gravitational potential implies the universal equivalence of inertial and gravitational mass. An alternate formulation is possible which parallels Einstein's original inductive reasoning from the universality of the constant of attraction to the geometrization of the law of gravitation
>
> (Kraichnan, 1956, p. 482)

In fact, at first glance, the property of the spin-2 to couple universally, regardless of its properties, seems to resonate well with WEP-like principles, which refer to the universal behaviour of concrete physical objects with arbitrary properties, and not so well with SEP-like principles, which refer to universal locality conditions on laws. So let me finally turn to the interpretation of universal coupling in the spin-2 approach.

Given the results of Sections 4.1 and 5.2.2, I distinguish the following versions of universal coupling for the contexts of the classical spin-2 theory (UC-C), and the quantum spin-2 theory (UC-Q), respectively:

UC-C   To obtain a consistent, Lorentz-invariant classical field theory, the classical spin-2 field h is required to couple universally to the total energy-momentum tensor.

UC-Q   To obtain consistent, Lorentz-invariant quantum particle interactions, gravitons are required to couple universally to all particle species including gravitons.[76]

For emphasis, both UC-C and UC-Q imply not only that the spin-2 couples to *other* fields in the same way, but to *all* fields including the spin-2 field. This means that gravitational energy and momentum are indistinguishable from energy and momentum of ordinary matter (e.g., electrons or photons) such that the source for the gravitational field is the total energy-momentum.

So, how do these universal coupling principles of the spin-2 theories and the traditional versions of the equivalence principles of standard general relativity relate? Do UC-C and UC-Q imply (one of) the traditional standard general relativity principles? In particular, do they imply the strongest general relativity principle SEP-GR? Or can we at least argue that the fact that a version of universal coupling holds for spin-2 theories explains the fact that (one of) the equivalence principles of general relativity hold(s)? As mentioned, physicists are quick to *identify* universal coupling (to the spin-2) with SEP-GR (i.e., "the" equivalence principle that entails general relativity). So how is this identification justified, if at all?

A natural idea is to first relate "universal coupling of the (quantum) spin-2 field" to "universal coupling of g", and then argue that SEP-GR follows. Note, however, that universal coupling of g might be viewed as just one of the requirements for SEP-GR – there may be additional requirements (similarly, see Ortín, 2017); for example, one might want to additionally exclude certain types of (non-dynamical) fixed fields from the theory (e.g., fixed timelike vector fields), as they can violate Lorentz invariance. Indeed, SEP-GR has been analysed as "universal plus minimal coupling" (to g) by Brown (2005); see endnote 75. However, Read, Brown, and Lehmkuhl show that this is not correct (Read et al., 2018); while universal coupling (to g) is necessary but not sufficient for SEP-GR, minimal coupling (to g) is a stronger constraint that, as they show, is neither sufficient nor necessary, but may violate certain versions of SEP-GR.

Another idea would be to argue that Weinberg's UC-Q is best identified with or implies WEP-GR, which seems in line with Weinberg's (1964a) own understanding, and then show that this plus another condition implies SEP-GR (WEP-GR does not suffice for constructing general relativity). For example, one could argue that SEP-GR follows from WEP-GR plus minimal coupling because all terms violating SEP-GR essentially behave as high-energy corrections and are therefore absent in the low-energy limit. Then the graviton's universal coupling could effectively imply SEP-GR. As discussed above, this is similar to Schiff's conjecture that WEP-GR implies other versions of the equivalence principle. In fact, we would hereby be able to clarify the plausibility argument that Schiff provides.

Roughly speaking, UC-C and UC-Q state in the terms of their respective theoretical frameworks that gravity does not distinguish *in any way* between different types of matter (or energy). Here, I use "matter" in the broadest sense: anything that may contribute to the energy-momentum tensor in a classical or quantum theory is considered as matter. In the quantum case, this includes, for example, electrons, quarks, and the particles that represent the fundamental laws (the photon, the graviton, gluons, and the Z^0 and $W^{\pm}$ bosons). The fact that, according to UC-C and UC-Q, the graviton couples to the energy-momentum tensor and material particles *regardless of their specific properties* suggests that UC-C and UC-Q are of the WEP-like type (see above). But the fact that, according to UC-C and UC-Q, the graviton couples *in the most general sense* to *any* type of energy-momentum (or, equivalently, *any* type of particle) in the same way – such that UC-C and UC-Q are not limited in scope – suggests that UC-C and UC-Q might imply (without argumentation à la Schiff) stronger statements than merely WEP-GR – potentially even SEP-GR. In particular, UC-Q's notion of "all particle species" is, on the face of it, far more general than WEP-GR's notion of a test body; also, UC-Q is about any fundamental process, while WEP-GR is only about processes of mechanical motion.

In a sense, Schiff-type reasoning suggests itself in this context. In fact, it is the quantum context that makes it perfectly transparent: here, the fundamental and, notably, innately local interactions are represented by point particles (e.g., photons) to which the spin-2 couples universally. So UC-Q implies a constraint condition for all fundamental laws: all fundamental interactions "experience" gravity in the same way. Local interaction experiments (including gravitational ones) cannot detect the presence of a background gravitational field. Moreover, since the symmetry properties of these interactions happen to be Lorentzian, the spin-2's universal coupling does indeed imply the strongest statement: SEP-GR. This argument is more obvious in the quantum particle case, but the result holds in the classical spin-2 case as well, since all interactions contribute to the energy-momentum tensor.

Based on the insights from the spin-2 theories, one can then show that the spin-2's universal coupling along with the fact that all physical fields are defined on η (or, put dynamically, have Lorentz-invariant dynamics) implies universal coupling for the g field of general relativity: g *inherits the universal coupling property from the spin-2*. The previously established assertion that a full-blown dynamical approach to spin-2 gravity is feasible, further bolsters this claim, as it justifies that η should be treated just as in the dynamical approach to special relativity.[77] So, UC-C and UC-Q imply g's universal coupling property and its local Lorentzian symmetry property in general relativity. Thus, both arguably imply SEP-GR.

However, there is a major problem with this argument. It starts from the assumption that all interaction dynamics are globally Lorentz-invariant, which, in a special-relativistic set-up, implies *local* Lorentz invariance. Consequently, one *assumes* rather than derives SEP-GR (which states that all laws are locally special-relativistic). Contrary to what is usually claimed in the physics literature (e.g., Nicolis, 2011), SEP-GR is a presupposition, not a result.[78] So what is Weinberg's argument good for?

First of all, Weinberg (and similarly Deser) explicitly demonstrate that SEP-GR is the essence of general relativity, since one can derive the Einstein field equations from demanding that all fundamental local interactions are Lorentz-invariant (as SEP-GR states). Second, the formal requirement of Lorentz invariance in a special-relativistic spin-2 theory certainly does not wear the specific physical implication for the coupling properties of the spin-2 on its sleeves; similarly, the formal requirement of Lorentz invariance in a special-relativistic spin-1 theory does not wear on its sleeves that the respective charge must be conserved. Only Weinberg's derivation reveals the physical content of these symmetry assumptions. This is a decisive advance. Reformulating the geometrically loaded content of SEP-GR is vital for understanding the new non-geometrical context. It removes the ambiguities of the different formulations of the equivalence principle and condenses them into a single principle. So, while we probably cannot uphold the view that Weinberg's argument is a true derivation of SEP-GR, it remains true that Weinberg *proves* that the physical content of the equivalence principle is about the fact that gravity couples universally to all fields. To fully appreciate this result, it is important to keep in mind that we are dealing with two entirely different theoretical frameworks here. The fact that Weinberg derives UC-Q from Lorentz invariance or reformulates SEP-GR as UC-Q *is* a significant result.

Furthermore, although the linear theory is manifestly Lorentz-invariant (also locally), such that the strong equivalence principle is a presupposition in the linear theory, we should *not* expect this to trivially translate to what holds in the full non-linear theory. Depending on one's analysis of the strong equivalence principle, non-linearity may spoil the formerly manifest strong equivalence principle. For example, minimal coupling (manifest for the spin-2 case with respect to η) will – in contrast to universal coupling – generally not be manifest for g in general relativity, because inheriting minimal coupling is not secured by the reduction: the other fields (which are part of the reduction basis) are not restricted to couple minimally to h, but only to η. After all, the fixed background η we started from is rendered unobservable (or rather operationally inaccessible) in the full non-linear theory. On the other hand, we are able to show that a reformulated version of the principle (i.e., universal coupling to g) *must* hold, since both non-linearity and universal coupling to h are forced on us in the construction; and, since g's universal coupling property only depends on the fact that h couples universally, this is preserved in the non-linear theory. So again, obtaining the reformulated version of the strong equivalence principle is not a trivial result.

Accordingly, only properties of the non-linear theory that are strictly reducible to what is manifest in the linear theory (e.g., universal coupling), will hold strictly in the full non-linear theory. Properties of the non-linear theory that are not strictly reducible to the properties manifest in the linear theory (e.g., minimal coupling) will generally not hold strictly, but at most effectively (e.g., in the low-energy regime) in the full non-linear theory – something we would expect when going from the linear to the non-linear theory. For low energies we obtain general relativity (possibly uniquely), while for high energies deviations from general relativity (modelled as high-energy corrections) are not constrained by the theory.

The graviton not only universally couples to any type of ordinary matter, but to gravitons as well. Moreover, the graviton's self-coupling property follows automatically, if any graviton–matter coupling is allowed. Accordingly, WEP-like statements in this context also encompass that "gravitational binding energy of massive bodies will not modify their free-fall motion" (Ortín, 2017, p. 11); Ortín dubs universal coupling a "microscopic version" (Ortín, 2017, p. 11) of the strong equivalence principle.

This emphasises that Weinberg's universal coupling result is indeed the most general or strongest reinterpretation of the equivalence principle there can be. Arguably, this fact that Weinberg's argument is internally forced to dismiss any restriction is why the physics literature takes it as a derivation of the strong equivalence principle. For example, Nicolis' (2011) assessment is consistent with this. After deriving the graviton's universal coupling to the (ordinary) matter energy-momentum tensor, Nicolis speaks of "the equivalence principle" (I assume that he refers to the *weak* equivalence principle, here). After noticing that self-coupling needs to be included for consistency, and thus to obtain coupling to the total energy-momentum tensor, Nicolis explicitly refers to it as the *strong* equivalence principle.

Let me also briefly comment on why such reformulations are to be expected, especially for quantum-theoretical treatments. Here is what hinders drawing the direct connection from Galileo's original version to the quantum context:

> The Galilean form of the equivalence principle – essentially that all bodies follow the same trajectory when freely falling in a gravitational field with the same initial conditions – is obviously inappropriate when applied to individual leptons. Quantum particles do not follow well defined classical trajectories, and even repeated experiments with a single electron would fail to duplicate the results because of quantum uncertainty.
>
> ...
>
> Faced with this spread in results caused by quantum uncertainty, the best one could do is to define some sort of average trajectory based on expectation values. But even in this quasi-classical sense, a straightforward statement of the principle of equivalence encounters a major obstacle. Fermions, such as electrons, do not follow geodesic paths in a gravitational field, because of the well known spin-curvature coupling .... The spin effect is mass-dependent, which implies that initially coincident electron and muon trajectories (even quasi-classical) will gradually diverge. ... Evidently the most elementary known particles actually fail to comply, even in a statistical sense, with the traditional statement of the equivalence principle.
>
> (Davies & Fang, 1982, p. 470)

Thus, it can be anticipated that in an explicitly non-geometrical reading of classical field theory and in the manifestly non-geometrical quantum context, we will be confronted with the need to reformulate the physical content of Galileo's experimental results or Einstein's elevator thought experiment, which are at the heart of the geometrical reading of general relativity. This, Davies and Fang continue, then "prompts the search for a more fundamental statement that is directly applicable to fundamental particles. ... A possible statement of the equivalence principle might then be that all fields couple with equal strength to gravity" (Davies & Fang, 1982, p. 471). This understanding is very much in line with my reformulation claim.

To summarise, Weinberg does not derive SEP-GR from more fundamental assumptions, but derives from SEP-GR (analysed as Lorentz invariance) a non-geometrical reformulation of the physical content of SEP-GR that applies to (quantum-)field-theoretic contexts. Universal coupling may have been a way to think about the strong equivalence principle all along, but in the spin-2 approach this particular reformulation is properly derived, made precise, and shown to translate to general relativity.

5.2.4   Reduction of General Relativity to Quantum Field Theory[79]

Now, as soon as the spin-2's universal coupling is established, the story is basically the same as in the classical case. In particular, the full non-linear classical theory can be constructed from here as the classical limit.[80] Accordingly, similar philosophical observations apply. Given that the spin-2's universal coupling property (including self-coupling) and Einstein's field equations are already derived in classical spin-2 theory, one might suspect that the (notably, independent) special-relativistic quantum particle derivation merely rediscovers the classical result and confirms that the quantised theory also has this feature.[81] This arguably tells against what sometimes seems to be implied in the literature, namely that substantial insights from the spin-2 perspective are only to be expected in quantum field theory (e.g., Cao, 1998; Carlip, 2014; Carroll, 2004; Nicolis, 2011; Salimkhani, 2018). If, instead, we see quantum spin-2 theory as just the quantised version of classical spin-2 theory, we can explain its success as one by design, namely as parasitic on the success of general relativity (Salimkhani, 2020). However, given that relativistic quantum particle theory (or, indeed, quantum field theory) is arguably a more fundamental framework than classical field theory,[82] quantum spin-2 theory *goes beyond* classical spin-2 theory in its decisive support of an ontological reduction of g. So systematically, it is the quantum version that should be considered more important. Let me dwell on this a bit further.

In this perspective, the (more fundamental) quantum result is, ultimately, *the reason* for the result obtaining in the (higher-level) classical spin-2 theory – after all, h is supposed to be the classical limit of the graviton field. In particular, or so I claim, it is the quantum spin-2 theory that selects the preferred ontology, not standard general relativity or the classical spin-2 view (which, by itself, can be considered on a par with the standard general relativity view; see Section 4.3).

This can be supported by observing that the relativistic quantum framework seems also generally more restrictive than classical field theory; for example, due to the results of Weinberg's (1964b) low-energy theorem and the Coleman and Mandula (1967) no-go theorem, which are both based on a few general assumptions. In short, regarding what survives long-range, the relativistic quantum framework only allows for massless fields with spins 0, 1/2, 1, 3/2, and 2. If this is correct, the constraint stems from the special-relativistic quantum particle derivation[83] – which, as mentioned, naturally drops out of quantum field theory, but is not tied to quantum field theory, but more generic. Note also that the spin-2 particle is necessarily part of string theory, which arguably further bolsters the spin-2 view, and may present the most fundamental reason for said results.

Since general relativity is obtained independently as the classical limit of the corresponding quantum field theory, general relativity is arguably reduced to quantum field theory. Accordingly, general relativity is not a fundamental theory,[84] but

> an essentially phenomenological (albeit unique) theory for describing interactions at macroscopic distances and times.
> (Boulware & Deser, 1975, p. 230)

This underlines the foundational importance of the spin-2 approach. In terms of principles, general relativity can be deduced from bringing together special relativity and quantum mechanics:[85]

> All of these things that Einstein did – Einstein thought about these falling elevators and he discovered the principle of equivalence and all these deep facts about classical physics that led him to think about general relativity – all of those things could have been discovered by much more mediocre theoretical physicists who knew about quantum mechanics.
> (Arkani-Hamed, 2013)

Typically, the conviction is that general relativity is more fundamental than special relativity, but according to the analysis above, it is, in fact, the other way around.

In a recent paper, Lehmkuhl (2021) argues that the Einstein equivalence principle (roughly, EEP-GR) can be considered a bridge between general relativity and Newtonian theory, and the strong equivalence principle a bridge from general relativity to special relativity. In light of my presentation, we can amend this to say that – in some appropriate interpretation of universal coupling and SEP-GR – the strong equivalence principle indeed remains a bridge between general relativity and special relativity, but the direction is turned around: Lorentz invariance – i.e., special relativity – is where we start, while general relativity is what we end up with. The strong equivalence principle is a bridge in the sense that it is where the reduction from general relativity to a special-relativistic field theory provides explanatory insight. This is a particularly interesting result since the strong equivalence principle is usually perceived as closely tied to a geometrisation view of gravity, not a "field only" view. After all, even Brown (2005) employs the strong equivalence principle precisely to establish chronogeometricity. As I have argued, the strong equivalence principle additionally proves to be the "link" (i.e., bridge) from general relativity to (quantum-field-theoretic) quantum gravity which underwrites this further.

Lehmkuhl (2021) also seeks to understand the role of the principle, and how it constrains the theory. From the spin-2 perspective, I take it that we should answer that the focus regarding the conceptual foundations of general relativity shifts from Lorentz invariance in disguise (i.e., a geometrical version of the strong equivalence principle) back to the conceptual foundations of special relativity (i.e., Lorentz invariance proper).

It is important to stress once again that quantum spin-2 theory is an effective field theory (Donoghue, 1994), which means that the quantum spin-2 approach only yields the low-energy limit of some full-fledged theory of quantum gravity. This restriction is actually less severe than one might think. Boulware and Deser argue that the special-relativistic quantum particle derivation is "in no way restricted to "weak field" situations … but only problems involving slow variation compared to particle Compton wavelengths" (Boulware & Deser, 1975, p. 197). In fact, due to this generality, the neglect of higher-order interactions therefore requires a further justification, which they do also provide. In short, higher-order terms are negligible as long as quantum corrections to the lower-order vertices are as well (Boulware & Deser, 1975, p. 197). Similarly, Weinberg's (1965a) derivation of the Einstein field equations is not restricted to the soft particle case, but relies on the more robust Dyson-Feynman perturbation theory.

Still, if the resulting theory is only an effective theory, why should quantum spin-2 theory be considered to have ontological import? With Weinberg (1999), one can reply as follows. In the modern effective field theory view, *all* of our best physical theories are not straightforwardly fundamental. Hence, we should be honest about what physics can and does provide us with:

> I think that in regarding the standard model and general relativity as effective field theories we're simply balancing our checkbook and realizing that we perhaps didn't know as much as we thought we did, but this is the way the world is and now we're going to go on the next step and try to find an ultraviolet fixed point, or (much more likely) find entirely new physics.
>
> (Weinberg, 1999, p. 250)

But – *pace* Redhead (1999), for example – this does not mean that effective field theories thereby block the search for fundamental laws of nature or undermine scientific realism.[86] In particular, one should keep in mind that also classical general relativity is effective in that it does not contain higher powers of the curvature tensor.[87] So quantum spin-2 theory merely draws attention to the fact that general relativity is an effective field theory as well.

5.3　Fundamentality of Spacetime Revisited

To conclude, all results from the classical spin-2 approach translate to the quantum version, but it is the quantum spin-2 approach that finally breaks the "ontological tie" between general relativity and (classical) spin-2 theory: the fundamental ontology is determined by quantum spin-2 theory. A dynamically interpreted quantum spin-2 approach suggests an understanding of spacetime as an at least *partly* non-fundamental structure: *metrical* aspects of spacetime are derivative on the dynamics of quantum matter including gravitons.[88]

Since spin-2 theory ontologically reduces g to the underlying dynamics, it is not merely chronogeometricity and the universal symmetry properties of the metric which are explained. Rather, the matter field dynamics pass on *further* properties to g (unlike what I have implicitly assumed here so far). In particular, the matter field dynamics may equip the metric field with an *orientation*, if at least some of the matter fields violate charge conjugation parity symmetry (CP-symmetry), which implies the violation of time reversal symmetry. This could be of interest for conceptualising the direction of time.[89] The fact that, according to a properly understood dynamical approach, metric fields like η and g are ontologically reduced to matter field dynamics, and hence inherit *all* the properties of these dynamics (rather than just the *universally shared* symmetry properties, which the proponents of the dynamical approach usually speak about), also helps to counter the argument from common origin inference à la Balashov and Janssen (2003) and Norton (2008): while it is true that a fundamental metric field may explain why *all* fields *share* certain dynamics – e.g., a fundamental η may explain why matter field dynamics are universally Lorentzian – it can*not* explain other non-universal dynamical symmetry properties of the matter fields. Already Brown has occasionally pointed out that the spacetime fundamentalist should be surprised why some matter fields violate parity, for example:[90] according to the symmetry properties of the Minkowski metric (and the g field alike) this should not be the case. If η is a fundamental entity that determines the dynamics of matter fields, as the fundamentalist maintains, then why should the Minkowski metric only transfer one of its symmetries, namely Lorentz symmetry, but not, say, parity symmetry? Brown's answer is, of course, that the Minkowski metric does not determine anything about the matter fields, but is determined by the matter fields. In fact, Brown goes a step further: to Brown, the Minkowski metric does not have any other properties, it is a non-entity, a mere codification of Lorentz symmetry. But there is a subtle alternative: already turning around the ontological dependence relation between the Minkowski metric and the matter fields provides a sufficient explanation. The respective metric field (e.g., the Minkowski metric) inherits *all* properties of the matter field dynamics (notably, the dynamics, not the fields themselves). As a result, one only obtains the Minkowski metric, if all matter fields have Lorentzian dynamics *and* do not violate parity symmetry, for example. The proponent of the dynamical approach does not need to argue that the Minkowski metric is merely a codification of Lorentz invariance: if one wishes, one can understand the Minkowski metric as a *physically real* metric that is derivative on matter field dynamics, *if* there are no gravitons and parity-violating matter fields. Then, rather than as a non-entity, we can understand the Minkowski metric analogously to g (which is the correct metric when the fundamental ontology most notably includes a graviton field), namely as a *non-fundamental entity*.

Now, what about *topological* aspects of spacetime, i.e., the manifold? Once again, the problem of pregeometry remains acute; and, once again, the possible solutions are the same as the ones presented in Section 3.6. Depending on which solution to the problem of pregeometry is adopted (constructivism or a half-way constructivism), the quantum spin-2 approach then arguably either supports some type of relationalism or a form of spacetime emergentism in the sense of Martens (2019).

Note also that spatiotemporality remains part of a Lorentzian theory like quantum spin-2 theory in a minimal sense, namely in terms of the space–time split. As many other theories of quantum gravity have this feature as well,[91] the space–time split may actually turn out to be fundamental (suggesting spacetime emergentism).

Notably, traditional spacetime fundamentalism *is not an option anymore*. This is because the operationally accessible spacetime – or rather, the operationally accessible metrical aspect of spacetime (i.e., the g field) – is unambiguously derivative on interacting quantum matter and not *vice versa* (unlike in the classical case): the direction of the ontological dependence relation, namely "g ontologically depends on gravitons (and not *vice versa*)", is firmly fixed. In particular, contrary to the classical case, i.e., *due to the quantum nature of the graviton*, the spacetime fundamentalist can*not* (1) simply insist on g being fundamental anyway, or (2) interpret what replaces g as spacetime. Recall that these *were* options in the classical case because $g = \eta + h$ did not specify a preferred direction of the ontological dependence relation between the three classical fields g, η, and *h*. At most, the spacetime fundamentalist can maintain – contrary to the proponent of a dynamical approach – that spacetime remains fundamental in the form of the Minkowski metric η. However, η is *operationally inaccessible*. Moreover, as we are in a quantum context, there is an additional argument that the proponent of a dynamical approach can bring forward: the Heisenberg collapse argument suggests that the interpretation of η as a mere codification of Lorentz symmetry is preferred due to the fact that the Heisenberg collapse argument tells against any form of fundamental metric structure in a quantum context (see Section 5.1.2).[92] Hence, the dynamical approach to quantum spin-2 theory, regardless of whether it is interpreted in terms of relationalism or spacetime emergentism, is clearly the more appropriate metaphysical position; again, especially because, in contrast to the classical case, here the Heisenberg collapse argument provides an additional argument against clinging to the Minkowski metric as fundamental. Relatedly, it is only the dynamical view of quantum spin-2 theory that brings out in full that the theory is a proper *quantum* theory of gravity, rather than a semi-classical theory: on a non-dynamical reading, the fundamental ontology does include the quantum fields *and* a classical Minkowski metric.

In fact, there is a notable difference between, on the one hand, classical spin-2 theory and general relativity, and, on the other hand, quantum spin-2 theory, which is, in principle, even empirically testable:

> [t]he spin-two meson theory of gravity [i.e., the spin-2 theory of gravity; my remark] differs from general relativity in that the meson theory is consistent with the uncertainty principle. If general relativity were the correct theory then the uncertainty principle would fail for large distances and low velocities; i.e., just in the classical limit. Thus the main purpose of a quantized theory of gravity is that it demonstrates the possibility of maintaining the uncertainty principle even to the classical limit.
>
> (Huggins, 1962, p. 1)

Accordingly, quantum spin-2 theory does *not* suffer from the same problems as classical spin-2 theory, which, as discussed, can be understood as a mere change of representation, whose ontological import is unclear.

However, as I have argued in Section 5.1.4, *all* theories and tenable interpretations of these theories that are compatible with the available empirical data are relevant for determining ontological inferences; they determine the metaphysical status of spacetime negatively by determining what is ruled out for not meeting a *common core* constraint. Accordingly, the geometrical interpretations of classical general relativity and classical spin-2 gravity are still pushing against the inference that metrical aspects of spacetime are non-fundamental.

In what follows, I attempt to resolve this issue by taking into account physics unificatory practise. In a nutshell, the idea is to constrain the ontological inferences in the philosophy of spacetime by means of continuity conditions to other parts of physics, in particular, fundamental particle physics. It is such conditions, or so I argue, that provide criteria for which of the various ontological commitments suggested by considerations in the philosophy of spacetime are ultimately preferred, namely for not only being suitable within a restricted domain of physics, but for respecting and being compatible with other central insights of physics as well.[93] To put it concretely, spacetime fundamentalism may be a tenable position when only considering a certain class of spacetime theories, but untenable when including, for example, insights from quantum physics.

The fact that only a sufficiently large scope can robustly determine the fundamental ontology is already suggested by how we standardly understand the term, namely as the minimal complete set of ontologically independent entities (see Chapter 2). To robustly determine what is in this set, it is essential to test the alleged ontological independence of the candidate entities – in principle, against the whole context of physics theory. Note that this does not tell against the idea that metaphysical claims are subject to constant revision: *new* empirical and theoretical insights may of course still lead to changes.

For such reasoning to be convincing, i.e., for preventing metaphysical inferences from being "philosophically engineered", it is crucial to show that unification is not some "philosophical ideal", i.e., that physics' engaging in unification is not due to a philosophically motivated constraint on physics practise. Therefore, I attempt to show in the following that instances of unification are, in fact, the *result* – or, for dialectical effect: the *by-product* – of concrete physical research that is committed to empirical adequacy and theoretical consistency.

Notes

1 For a nice historical overview of the development of quantum gravity see Rickles (2020).
2 The issue is sometimes misrepresented as follows: that we do not know how to combine general relativity, which is argued to be a theory of very large scales, and quantum mechanics, which is argued to be a theory of very small scales (e.g., Huggett and Callender, 2001b). Here is a sense in which this is misleading: for small energies (i.e., large spatiotemporal distances), combining general relativity and quantum mechanics does not pose a problem. For small energies, quantum gravity does not pose a problem. All the issues that set off the whole research programme of quantum gravity and its diverse theory proposals arise for high energies where gravity becomes the dominating force such that a straightforward quantisation of general relativity is not predictive anymore. In fact, it is possible to calculate the (negligible) effect of a quantum-mechanical correction to Newton's law of gravitation (Arkani-Hamed, 2010b):

$$F = \frac{Gm_1m_2}{r^2}\left\{1 - \frac{27Gh}{2\pi^2r^2c^2} + ...\right\}$$

Similarly, a low-energy theory of quantum gravity is already at our disposal (see Section 5.2).
3 Generally, the "quantisation" of some classical theory (e.g., general relativity) means the construction of a quantum theory whose classical limit agrees

with the classical theory. There are different methods of quantisation. Roughly speaking, the classical field variables are converted to operators. Note that quantisation does not necessarily imply discreteness (i.e., that the eigenvalues of the operators are discrete). Consider the case of quantum mechanics: the eigenvalues of some observables are always discrete (e.g., spin and angular momentum), but others may generally have continuous eigenvalues (e.g., position and momentum); here, discreteness can then arise through boundary conditions. Accordingly, the quantisation of general relativity does not imply discretisation of spacetime.

4 See Crowther and Linnemann (2019).

5 For a philosophical assessment of the context of discovery regarding quantum gravity research see Linnemann (2019).

6 For example, Huggett and Wüthrich (2013), Matsubara (2017), and Martens (2019).

7 For example, Paul (2012). Paul does not primarily draw her conclusions from quantum gravity, though. Instead, Paul (2012) argues against what she dubs "spatiotemporalism" – i.e., the view that the fundamental constituents of the world are "chunks of spacetime" – based on so-called configuration space realist (or wave function realist) interpretations of quantum mechanics (see Ney, 2021). Moreover, Jaksland (2021) proposes entanglement as an alternative to spatiotemporal distance as a world-making relation.

8 For example, Le Bihan (2018; 2021), Wüthrich (2018), Matarese (2019), Baron (2020; 2021), and Baron and Le Bihan (2023).

9 On large scales, gravity is extremely weak compared to the other interactions. Suppose we take two macroscopically separated electrons and compare the magnitudes of their gravitational attraction and their electromagnetic repulsion. We find that the gravitational force is considerably smaller, namely by roughly 40 orders of magnitude. Now, one could think that this ratio remains constant independent of the distance between the electrons, since both the gravitational and the electromagnetic force are governed by an inverse-square law. However, while the electromagnetic coupling constant $\alpha \approx 1/137$ is dimensionless, the gravitational constant $G_N \approx (10^{-33} \text{cm})^2$ has a dimension. The strength of the gravitational interaction is therefore scale-dependent.

10 There is a sense in which the so-called Planck scale is the highest energy scale. It refers to energies of the order of 10^{19} TeV. Quantum-gravitational effects are expected at around 10^{17} TeV. For comparison, the currently highest energy scale probed (by the Large Hadron collider) is roughly 14 TeV. Thus, the typical energy scale where quantum-gravitational effects are understood to become relevant is roughly 16 orders of magnitude higher than presently experimentally accessible (e.g., Arkani-Hamed, 2012).

11 For more details, see Carlip (2001), Kiefer (2006; 2007), Huggett and Callender (2001b), and Wüthrich (2005). The "appeal to the unity of physics" is what Carlip (2001, p. 888) dubs the "stock reply". In Chapter 6, I argue that this understanding of unification is putting the cart before the horse: physics does not assume unity, but ends up with it.

12 Note, however, that despite the fact that quantum-gravitational effects are primarily higher energy effects, they might still show as tiny deviations from the current theory's predictions at accessible energy scales. Furthermore, there are proposals for how to assess theories in the absence of empirical data; see Dawid (2013), Chall (2018), Dardashti et al. (2019), and Menon (2019).

13 Given the work of Section 3.5, it is interesting to note that being wrong about matter results in a conceptual issue for spacetime as well – at least in the perspective of the dynamical approach. This is because there is a sense in which general relativity, as it stands, is not like the other classical field theories: the dynamical approach argues that general relativity (and special relativity, respectively) ultimately cannot be understood conceptually when isolated from the rest of physics. It is the non-gravitational quantum fields, which build rods and clocks, that are supposed to convey chronogeometricity to the metric field. But in general relativity these non-gravitational fields are conceived of as classical. Therefore, strictly speaking, the dynamical approach's key insight regarding chronogeometricity is not consistently resolved within general relativity itself. There is a sense in which general relativity as such conceptually *depends* on quantum theory, but still treats everything classically. Hence, "there is a deep-seated tension in the story about how the metric field gains its chronometric operational status" (Brown, 2005, p. 10): "if the Minkowski space-time interval is to have some connection with the universal behaviour of rods and clocks ... ultimately some appeal to quantum theory must be made. In SR [special relativity], this raises no problem in principle: the big principle after all does not pin down the precise form of the dynamics of matter. But the situation is more complicated in GR [general relativity]. There is, after all, a crucial tension between the dynamical interpretation of SR and the structure of GR. It arises from the fact that Einstein's field equations refer to classical fields. In particular, the matter fields appearing in the definition of the stress-energy tensor are classical" (Brown, 2005, p. 177). We might take this as a conceptual hint from general relativity that we should – at least – equip general relativity with quantum matter fields. Given the work of Chapter 4, however, we might as well point out that according to general relativity, spacetime (or, more precisely, metrical aspects of spacetime) is already derivative from matter fields.

14 For critical assessments and responses to Eppley and Hannah (1977), see Huggett and Callender (2001a; 2001b), Mattingly (2005; 2006) and Wüthrich (2005).

15 See Arkani-Hamed (2010b), recently referred to also by Martens (2019). For a comprehensive review of such arguments, see Hossenfelder (2013).

16 Because of this property of gravity, nothing hinges on the reformulation from momentum to energy.

17 The variant of the argument due to Doplicher et al. (1995) defines that spacetime only has an operational meaning if measuring signals can leave the region in question. Wüthrich (2005) points out that the argument is independent of this definition.

18 Doplicher et al. (1995) then continue by investigating "how localization is restricted by requiring that no black hole is produced in the course of measurement" and propose uncertainty relations for the different coordinates of spacetime events.

19 See also Martens (2019).

20 See Dowker et al. (2004).

21 According to Martens (2019, p. 6), the argument is originally due to Wheeler in Misner et al. (1973).

22 Huggett (2017) advances a similar line of thought when arguing in analogy to the possibility of preferred rest frames in special relativity that the target space does not have a determinate radius in string theory – and therefore should not be identified as phenomenal spacetime: "one could claim that there is a

preferred rest frame in spacetime, even though it has no physical influence in special relativity. One could even claim that it is some frame which can be picked out physically and phenomenally: for example, perhaps the fixed stars (idealised as an inertial frame) are at rest. I think that the proposal will strike most people as completely unmotivated. But replace 'frame' with 'radius', and the fixed stars with the phenomenal radius, and the parallel is perfect. Looked at this way, the definite radius view appears as a reactionary attempt to preserve aspects of an old theory when it is superseded, and understood as merely effective." (Huggett, 2017, p. 87)

23 At familiar scales there are objects we call "tables". The question then is whether tables are fundamental and hence should be included in our list of fundamental objects, or whether tables ontologically depend on some underlying structure that, in turn, does actually make it into the list of the fundamental ontology. To advance this argument, one needs to argue why tables *at the scale where they are empirically adequate* are nevertheless non-fundamental (because they are large-scale aggregations of atoms).

24 In fact, I take it that the independence notion of fundamentality does *imply* a sense in which the fundamentals cannot be inconsistent. This is because "fundamental facts should be *freely modally recombinable* (Armstrong 1997, 196; Bennett 2011, fn. 6)" (Muñoz, 2020, p. 212). Since the fundamentals do not depend on each other, they can be combined without restriction, which implies that they cannot contradict each other. So, for any combination of the fundamentals, we should be able to find a possible world in which these represent the complete set of fundamentals (Muñoz, 2020, p. 212). Note that in physical theories (or interpretations thereof) non-fundamentality as dependence might therefore occasionally show up as theoretical inconsistency. If a theory or interpretation builds on entities that were assumed as independent and freely modally recombinable, but are not, this may lead to inconsistencies. Concretely, assuming gravitation, matter, the Heisenberg uncertainty principle, *and* a fundamental continuous spacetime that posits arbitrarily short distances is inconsistent because what the first three fundamental ingredients entail contradicts the latter. The four ingredients are not freely modally recombinable *because* they are dependent.

25 This raises the question whether there are any constraints on what can serve as a fundamental. In particular, can only positive or actual entities ground, or should we allow for grounding by absence? Bernstein (2016, pp. 26–27) argues that absences can figure as causes, but not as grounds. Baron and Le Bihan (2023) have recently used grounding by absence in their causal theory of spacetime.

26 Still, semi-classical gravity is arguably an effective theory that is useful for specific systems where quantum effects are studied in strong gravitational fields. For example, semi-classical models become important for studying the Hawking radiation of black holes (see Hawking, 1975).

27 Notably, Mattingly (2005) disagrees with this assessment. He groups into three classes what physicists and philosophers bring forward as reasons to dismiss semi-classical proposals (and seek a full-fledged theory of quantum gravity instead – which for him is "a theory of gravity that treats the metric itself as a quantum field" (Mattingly, 2005, p. 325)): experimental reasons (including both actually performed experiments and detailed thought experiments), theoretical reasons (e.g., divergency issues of the expectation value $\langle T_{\mu\nu} \rangle$ and unstable solutions of the modified Einstein field equations), and metatheoretical

reasons (in particular, some ideal of unification). Mattingly then argues that it is the last class of arguments "that is really responsible for the conviction (quite widespread in the physics community) that gravitation is necessarily a quantum mechanical phenomena" (Mattingly, 2005, p. 326). Note that it is a bit unfortunate that Mattingly (2005), Wüthrich (2005), and others as well argue about whether it is "necessary" or "contingent" that gravitation (or spacetime) is quantised. After all, the debate is not about questions of modality, but merely about assessing the arguments in favour and against gravity requiring some form of quantum-theoretical description in the actual world.

28 The argument only assumes that the gravitational interaction between two masses is mediated and not a direct interaction at a distance, and that two systems cannot be entangled by local operations and classical communication (Bose et al., 2017).

29 See Huggett and Wüthrich (2013) for a slightly more comprehensive and partly more detailed survey.

30 See, for example, Kiefer (2007, p. 43).

31 This is the effective field theory programme. See Donoghue (1994). For more details on effective field theories see Georgi (1993).

32 As a side note, see Duff (1974) for a nice discussion of whether the perihelion precession of planetary orbits is a linear or non-linear effect.

33 See Salimkhani (2018).

34 See Niedermaier and Reuter (2006) for an extensive review of asymptotic safety and Friederich (2018) for a philosophical appraisal.

35 Notably, this issue is circumvented in string theory due to the conjectured gauge/gravity duality (i.e., the anti-de Sitter/conformal field theory (AdS/CFT) correspondence) where gravity is a quantum field theory not in d but $d-1$ dimensions (Shomer, 2007).

36 See Vistarini (2019) for a detailed study of spacetime in string theory and Rickles (2014) for a historical overview.

37 String theory can only consistently describe both bosonic and fermionic quantum particles in ten dimensions. As Dawid (2013, p. 12) points out, "[t]his prediction marks the first time in the history of physics that the number of spatial dimensions can be derived in a physical theory".

38 The non-renormalisability issue arises in a Hamiltonian formulation in terms of a non-closed Hamiltonian constraint algebra. "Attempting to close the constraint algebra by adding the 'right hand side' as an additional constraint just leads to new terms and so on" (Shomer, 2007, p. 8).

39 Wüthrich urges caution, however: "to read off the Wheeler-DeWitt equation that there is no time at all, however, is a bit too quick. It may still be there, of course, but in a way such that the dynamics may not be wearing it on its sleeves" (Wüthrich, 2012a, p. 9).

40 This is different from similar approaches (e.g., causal dynamical triangulation; see Loll, 2019) where discretisation is intimately connected to a quantisation attempt.

41 Note that causal set theory does not give rise to the full solution space of general relativity. For example, due to the properties of the causal precedence relation, causal set theory cannot accommodate general-relativistic spacetimes with closed timelike curves.

42 The process of distributing the points is called "sprinkling". There is an issue with determining whether the causal set structure one starts from allows for the emergence of a Lorentzian spacetime. To circumvent this, one can reverse-engineer

the desired class of sets by taking a Lorentzian manifold and "sprinkling" points on this manifold. Lorentz invariance can be retained if the sprinkling is performed according to a random Poisson process (Dowker et al., 2004).

43 See also Jaksland and Salimkhani (2023).

44 Huggett (2017) argues that the significant difference between phenomenal spacetime and target space regarding the degree of spatiotemporality is that "'target space' is not a 'space' in the familiar sense at all" (p. 88), but a space equipped only with the structures that both T-dual theories have. Hence, it does not have a determinate radius.

45 See Esfeld (2021).

46 There are more radical proposals. McKenzie (2022, pp. 55–65) even includes false physical theories in her indirect argument against foundationalism (i.e., the claim that fundamentality relations "bottom out into a set of ungrounded fundamentalia" McKenzie, 2022, p. 10). See also McKenzie (2011, 2017).

47 Due to the external symmetries of special relativity, quantum particles are classified by mass and spin values (see Section 4.1.1; see also Duff, 1975, p. 80). Additional internal (gauge) symmetries give a further Casimir invariant connected to a charge.

48 Depending on the starting point there are several related ways to arrive at the correct properties the graviton needs to have. For example, Carroll (2004, pp. 298–299) first considers gravitational wave solutions to linear Einstein theory and then argues that the rotational symmetry properties of the corresponding polarisation vectors yield that the graviton is a spin-2 particle.

49 Half-integer spin particles can only account for scattering, but not for a static force like gravity or electromagnetism (Hatfield, 1995, p. xxxiv). The reason is that a half-integer spin particle changes the internal state of the particle it is emitted by or absorbed from (Hatfield, 1995, p. xxxiv).

50 Notably, there is a fourth phenomenological property, namely that gravity obeys the (weak) equivalence principle, which roughly states that (borrowing Newtonian notions for a moment) every object in a gravitational field falls at the same rate regardless of its properties – in particular, regardless of its mass. It turns out, however, that this property is actually not fundamental, but derivative; hence, for what follows let us forget about this property and only come back to it at the end of Section 5.2.2 and in more detail in Section 5.2.3.

51 This also indicates that the representing tensor should be symmetric, as the anti-symmetric parts behave like spin-1 fields (Hatfield, 1995, p. xxxv).

52 See, for example, Hatfield (1995, pp. xxxiv–xxxv): a spin-0 graviton would only couple to the trace of the energy-momentum tensor. Since the electromagnetic energy-momentum tensor is traceless, a spin-0 graviton would therefore not couple to photons. As a result, photons would not be affected by gravity – *contra* the empirical findings. Hence, the graviton cannot have spin zero.

53 Orthography and punctuation in this quotation have been slightly modified.

54 For the details of the different original works see Weinberg (1964a, 1964b, 1965a, 1965b), Boulware and Deser (1975), and Davies and Fang (1982). The following presentation is a revised version of a previously published summary in Salimkhani (2018) and is reproduced with permission from Springer Nature. It rests on Weinberg's original work, on his textbook on quantum field theory (Weinberg, 1995, pp. 534–539), a lecture by Arkani-Hamed (2010a), and lecture notes by Nicolis (2011).

55 In principle, such a scattering amplitude may be determined using the Feynman rules, which can be read from the corresponding Lagrangian. In particular,

considering the full details of the scattering process would mean considering all possible Feynman diagrams for the process $\alpha \to \beta$.

56 To obtain expression (5.3), it is used that the particle is *on shell* such that $p_i^2 = M_i^2$. Recall that according to special relativity a particle has four-momentum $P_\mu = (E, \vec{p}c)$ where E is the particle's energy, $\vec{p}$ is the particle's three-momentum, and $P^2 = m_0^2 c^4$, where m_0 is the particle's rest mass. This gives that $P^2 = P^\mu P_\mu = E^2 - \left|\vec{p}\right|^2 c^2 = m_0^2 c^4$ (note that the speed of light is standardly set to $c = 1$). The solution space of this equation forms a hyperboloid surface, the "mass shell". "On shell" or "real" particles satisfy the equation, "off shell" or "virtual" particles do not. For example, the photon in Figure 5.3 is off shell.

57 $\frac{1}{(p_i+q)^2 - M_i^2} \to \infty$ for $q \to 0$.

58 As Arkani-Hamed (2010a) emphasises, expression (5.2) can only become large, if the emission takes place on one of the incoming or outgoing states, which are on shell. Inside the grey circle in Figure 5.4b everything is off shell (see endnote 56) such that the propagator (5.2) does not reduce to the diverging term (5.3). Hence, ignoring the details of the scattering is justified. In particular, ignoring the possibility of soft emissions from within the grey circle is justified – such emissions do not contribute to the leading-order process.

59 This is the standard procedure of summing over all possible processes, since we cannot know which particle the photon is emitted from.

60 Here, the charge of a particle is defined as its coupling constant for emission of soft photons (Weinberg, 1965a, p. B989).

61 This means that we demand that the amplitude is invariant under transformations of the polarisation vector of the form $\varepsilon_\mu(p) \to (\Lambda\varepsilon)_\mu(p) + \alpha(\Lambda p)_\mu$ (see also Appendix A.2).

62 This means that we demand that the amplitude is invariant under Lorentz transformations; in particular, we demand invariance for $\varepsilon_{\mu\nu}(p) \to (\Lambda\Lambda\varepsilon)_{\mu\nu}(p) + (\Lambda p)_\mu(\Lambda\alpha)_\nu$ (see also Appendix A.2).

63 Nicolis (2011) additionally studies the implications of having two (or more) subsystems A and B of particle species that do not interact with each other. He shows that such subsystems can have different gravitational coupling constants κ_A and κ_B. However, if κ_A and κ_B are non-zero, i.e., if both subsystems interact with gravity, then it follows that κ_A equals κ_B – the gravitational interaction is, in fact, universal. If one of the subsystems, say, A, did not interact with gravity ($\kappa_A = 0$), it would follow that this subsystem is entirely isolated, i.e., we would have, as a matter of principle, no empirical access to it at all – which would arguably justify the use of Occam's razor. This underlines the robustness of Weinberg's result.

64 The all-important feature behind Weinberg's argument is that the degrees of freedom of massless particles with spin $s \geq 1$ are fixed by their two helicity states, which significantly constrains their long-range behaviour (see Arkani-Hamed (2010a) and Appendix A.2).

65 See Weinberg (1965a; 1965b; 1995) and Boulware and Deser (1975).

66 This section is based on work previously published in Salimkhani (2020). The text has been substantially revised and is reprinted with permission from Elsevier.

67 For systematic, experimental, and historical reviews, see, for example, Di Casola et al. (2015), Ghins and Budden (2001), Lehmkuhl (2021), and Will (2001).

68 There are other, non-traditional versions, and there are other formulations of the three traditional versions. For instance, Di Casola et al. (2015, p. 41) define

SEP-GR as: "[a]ll test fundamental physics (including gravitational physics) is not affected, locally, by the presence of a gravitational field"; and give two formulations of a weak equivalence principle, the latter of which is closest to WEP-GR: (1) "[t]est particles with negligible self-gravity behave, in a gravitational field, independently of their properties" (Di Casola et al., 2015, p. 40), dubbed WEP, and (2) "[t]est particles behave, in a gravitational field and in vacuum, independently of their properties" (Di Casola et al., 2015, p. 40), dubbed gravitational weak equivalence principle (GWEP). For some purposes, these formulations may be more useful, but it is non-trivial to see whether and why the different formulations have the same physical content. The different formulations usually refer to different theoretical and operational concepts. Therefore, their physical content is generally not exactly the same. The different versions and formulations of the principle may help to clarify empirical differences between the various theories of gravity. For instance, it is often argued that the strong equivalence principle is only satisfied by general relativity; see Di Casola et al. (2015).

69 The geodesic principle states that "free massive point particles traverse timelike geodesics", which we can regard as a relativistic version of Newton's first law of motion (Malament, 2012, p. 245).

70 Recall that in the Newtonian context the weak equivalence principle implies that the inertial mass m_i and the gravitational mass m_g of any object are equal in value (which Di Casola et al. (2015) call the *Newtonian equivalence principle* (NEP)), such that what has previously been dubbed the 'gravitational charge' of an object, namely m_g/m_i, is the same for all objects (Carroll, 2004, p. 48). Note that the weak equivalence principle implies NEP, but not the other way around: "Only as long as the two kinds of masses enter in the equations of motion through their ratio m_i/m_g alone, as happens in Newton's theory, does NEP imply the universality of free-fall. If different sorts of combinations of m_i and m_g are allowed, then spurious instances of the masses crop up and generically the WEP [weak equivalence principle] fails" (Di Casola et al., 2015, p. 42).

71 Recall that we usually want the symmetries of the matter field dynamics to match the symmetries of spacetime and *vice versa* (Earman, 1989, p. 46). If this matching was violated, then either spacetime or the dynamics have "more structure". If spacetime has more structure than the dynamics, there are spacetime structures that are undetectable by the dynamics. If the dynamics has more structure than spacetime, the dynamics can measure spacetime structure that does not exist.

72 For example, it is often claimed that SEP-GR is only compatible with general relativity, whereas WEP-GR is also compatible with alternative theories (see Di Casola et al., 2015).

73 This is precisely what the proponent of the dynamical approach utilises for directing problem cases at the proponent of the geometrical approach. The problem cases make use of the fact that, say, Newtonian matter fields (i.e., matter fields that have Galilean-invariant dynamics) may contribute to the energy-momentum tensor, unless the strong equivalence principle is assumed (see Section 3.5).

74 For example, regarding questions like "What (sub-)class of material bodies do we consider?" or "What class of responses (e.g., change of mechanical motion, change of interaction cross section, etc.) do we consider?", and so on.

75 SEP-GR has been analysed in terms of such concepts in the literature. First, SEP-GR has been analysed in terms of *Lorentz invariance* or, more precisely,

Poincaré invariance (Read et al., 2018). Second, SEP-GR has been analysed as Lorentz invariance plus *minimal coupling* – which implies that matter fields do not couple to the curvature tensor or its contractions (Brown, 2005). Third, SEP-GR has been analysed in terms of *universal coupling* to g – i.e., "[t]he idea that a single field g interacts with all the nongravitational [we shall soon encounter another version of universal coupling without this restriction; my remark] fields in a unique manner" (Will, 2001, p. 17) – plus minimal coupling (e.g., Brown, 2005; Will, 2001). It is important to note that these three variants are not equivalent, as Read et al. (2018) show. Minimal coupling is an independent condition that can actually lead to violations of SEP-GR (Read et al., 2018). This is why Read et al. (2018) drop minimal coupling in favour of Lorentz invariance as the sole condition.

76 Technically, this means that the coupling of any particle (including the graviton) to gravitons involves the same coupling constant κ (see Section 5.2.2).

77 Note also that minimal coupling (regarding η) is manifest for all fields (including the spin-2 field) in both versions of the spin-2 theory because they are from the outset nothing but special-relativistic theories. Minimal coupling in the spin-2 context (i.e., with respect to η) is not an additional presupposition, because it is included in the assumption of (global) Lorentz invariance: "[r]ecall that in SR [special relativity], inertial frames are global, which implies that the curvature vanishes everywhere, and hence trivially makes no appearance in the laws of physical interactions" (Brown, 2005, pp. 170–171). Note that this is not the condition revoked by Read et al. (2018) in the context of *general* relativity.

78 In fact, Will (2001) maintains – though without clarification – that all derivations of general relativity (including the ones we considered) make "implicit use of the SEP [SEP-GR]" (Will, 2001, p. 77).

79 This section is based on work previously published in Salimkhani (2020). The text has been substantially revised and is reprinted with permission from Elsevier.

80 However, it is important to note that there are caveats to this assessment. First of all, the solution space is restricted. Besides that, the probably most important issue is that for such a reduction to be carried out in full, one would need to show that realistic classical fields can actually be constructed from gravitons (see Huggett & Wüthrich, book manuscript; see also Huggett & Vistarini, 2015). Although this should be possible in principle, it is not clear whether it is possible other than for the weak-field approximation (because the non-linear dynamics might destroy the coherence of the state).

81 With respect to unification, it is certainly a philosophically interesting observation that general relativity can be constructed from quantum field theory independently.

82 Quantum physics is typically perceived as more fundamental than classical physics in the sense of the quantum typically being the "parts" of a classical whole. This can be made more precise by pointing out that the classical regime is a specific limit with respect to the underlying quantum regime: there are many more states on the quantum side (gravitons) than there are on the classical side (spacetime); for example, only coherent spin-2 states will yield general-relativistic spacetime. Moreover, as I argue in Chapter 6, issues of fundamentality are generally settled by unification, which is where quantum theories are clearly more relevant than classical theories.

83 Weinberg seems to hold this view: "In my opinion the answer is not to be found in the realm of classical physics, and certainly not in Riemannian geometry,

but in the constraints imposed by the quantum theory of gravitation" (Weinberg, 1972, p. viii).

84 Hence, Lehmkuhl's (2019) classification of general relativity as a "hybrid theory" is called into question: general relativity is fundamental neither regarding matter (which is uncontroversial), nor regarding gravity. As is well known, one can distinguish between "principle" and "constructive" theories. A constructive theory is usually understood to have higher explanatory power than a principle theory (see Brown & Pooley, 2006), as it draws on fundamental structures only. Similarly to special relativity, general relativity can be viewed as a principle theory, at least in part, although this is controversial. Lehmkuhl dubs it a "hybrid theory": fundamental regarding gravity, effective and phenomenological (and hence – or so I shall take it – more like a principle theory) regarding matter. There is a sense in which the theory is based on a number of postulated principles – that is, according to Einstein (1914), precisely formulated empirical facts. In particular, the key principle for the strong connection between gravity and spacetime in standard general relativity is the equivalence principle (e.g., Poisson & Will, 2014).

85 It is sometimes argued that another assumption is the "cluster decomposition principle" (Weinberg, 1999), which basically excludes that distant events are relevant for local experiments.

86 See also Williams (2019).

87 So one could argue that the most important reductive aspect of the spin-2 proposal concerns the reduction of a *theoretical concept* – the strong equivalence principle – to another (arguably more fundamental) theoretical concept – (global) Lorentz invariance. It is this conceptual reduction that – in light of the dynamical approach – then fuels the reduction in terms of ontology (cf. Morrison (2006) claiming that physics is usually about reduction of theoretical principles, not ontological entities).

88 Whether the graviton itself is fundamental is another issue that is famously addressed by Weinberg and Witten (1980), who give a concise no-go argument for why the graviton needs to be a fundamental particle. Loebbert (2008) evaluates the argument.

89 See Salimkhani (2022).

90 I take this from a presentation of Brown in Tübingen in August 2017.

91 See Le Bihan and Linnemann (2019) and Linnemann (2021).

92 The related problem of the theory's non-renormalisability remains acute, of course.

93 Similar to Haack's (1993) crossword analogy.

References

Abbott, B. P. et al. (2016). Observation of gravitational waves from a binary black hole merger. *Physical Review Letters*, *116(6)*, 061102. doi:10.1103/PhysRevLett.116.061102

Adlam, E. (2022). Tabletop experiments for quantum gravity are also tests of the interpretation of quantum mechanics. *Foundations of Physics*, *52*, 115. doi:10.1007/s10701-022-00636-z

Álvarez, E. (1989). Quantum gravity: An introduction to some recent results. *Reviews of Modern Physics*, *61(3)*, 561–604. doi:10.1103/RevModPhys.61.561

Arkani-Hamed, N. (2010a). Robustness of GR. Attempts to Modify Gravity. Part I. Prospects in Theoretical Physics Program: Frontiers of Physics in

Cosmology, Institute for Advanced Study, 25–28 July 2011. https://video.ias.edu/pitp-2011-arkani-hamed1

Arkani-Hamed, N. (2010b). The future of fundamental physics. Lecture 3: Space-time is doomed. What replaces it? Messenger Lecture Series, Cornell University, 4–8 October 2010. https://www.cornell.edu/video/nima-arkani-hamed-spacetime-is-doomed

Arkani-Hamed, N. (2012). The future of fundamental physics. *Dædalus, 141(3)*, 53–66. doi:10.1162/DAED_a_00161

Arkani-Hamed, N. (2013). Philosophy of Fundamental Physics. Part I. Andrew D. White Professors-at-Large Program. Cornell University, 1/3 October 2013. https://www.cornell.edu/video/nima-arkani-hamed-philosophy-of-fundamental-physics-1

Armstrong, D. M. (1997). *A World of States of Affairs*. Cambridge: Cambridge University Press.

Balashov, Y., & Janssen, M. (2003). Presentism and relativity. *The British Journal for the Philosophy of Science, 54(2)*, 327–346. doi:10.1093/bjps/54.2.327

Baron, S. (2020). The curious case of spacetime emergence. *Philosophical Studies, 177*, 2207–2226. doi:10.1007/s11098-019-01306-z

Baron, S. (2021). Empirical incoherence and double functionalism. *Synthese, 199*(Suppl 2), 413–439. doi:10.1007/s11229-019-02462-9

Baron, S., & Le Bihan, B. (2023). Causal theories of spacetime. *Noûs*, 1–23. doi:10.1111/nous.12449

Bennett, K. (2011). By our bootstraps. *Philosophical Perspectives, 25(1)*, 27–41. doi:10.1111/j.1520-8583.2011.00207.x

Bernstein, S. (2016). Grounding is not causation. *Philosophical Perspectives, 30*, 21–38. doi:10.1111/phpe.12074

Bose, S., Mazumdar, A., Morley, G. W., Ulbricht, H., Toro, M., Paternostro, M., Geraci, A. A., Barker, P. F., Kim, M. S., & Milburn, G. (2017). Spin entanglement witness for quantum gravity. *Physical Review Letters, 119(24)*, 240401. doi:10.1103/PhysRevLett.119.240401

Boulware, D. G., & Deser, S. (1975). Classical general relativity derived from quantum gravity. *Annals of Physics, 89*, 193–240. doi:10.1016/0003-4916(75)90302-4

Brown, H. R. (2005). *Physical Relativity: Space-Time Structure from a Dynamical Perspective*. Oxford: Oxford University Press.

Brown, H. R., & Pooley, O. (2006). Minkowski Space-Time: A Glorious Non-Entity. In D. Dieks (Ed.), *The Ontology of Spacetime. Philosophy and Foundations of Physics, Vol. 1* (pp. 67–89). Amsterdam: Elsevier.

Cao, T. Y. (1998). *Conceptual Developments of 20th Century Field Theories*. Cambridge: Cambridge University Press.

Carlip, S. (2001). Quantum gravity: A progress report. *Reports on Progress in Physics, 64*, 885–942. doi:10.1088/0034-4885/64/8/301

Carlip, S. (2008). Is quantum gravity necessary? *Classical and Quantum Gravity, 25*, 154010. doi:10.1088/0264-9381/25/15/154010

Carlip, S. (2014). Challenges for emergent gravity. *Studies in History and Philosophy of Science Part B: Studies in History and Philosophy of Modern Physics, 46*, 200–208. doi:10.1016/j.shpsb.2012.11.002

Carroll, S. (2004). *Spacetime and Geometry: An Introduction to General Relativity*. San Francisco: Addison Wesley.

Chall, C. (2018). Doubts for Dawid's non-empirical theory assessment. *Studies in History and Philosophy of Science Part B: Studies in History and Philosophy of Modern Physics*, *63*, 128–135. doi:10.1016/j.shpsb.2018.01.004

Ciufolini, I., & Wheeler, J. A. (1995). *Gravitation and Inertia*. Princeton, NJ: Princeton University Press.

Coleman, S., & Mandula, J. (1967). All possible symmetries of the s matrix. *Physical Review*, *159(5)*, 1251–1256. doi:10.1103/PhysRev.159.1251

Crowther, K., & Linnemann, N. (2019). Renormalizability, fundamentality, and a final theory: The role of UV-completion in the search for quantum gravity. *The British Journal for Philosophy of Science*, *70(2)*, 377–406. doi:10.1093/bjps/axx052

Dardashti, R., Dawid, R., & Thébault, K. (2019). *Why Trust a Theory?: Epistemology of Fundamental Physics*. Cambridge: Cambridge University Press.

Davies, P. C., & Fang, J. (1982). Quantum theory and the equivalence principle. *Proceedings of the Royal Society of London A*, *381*, 469–478. doi:10.1098/rspa.1982.0084

Dawid, R. (2013). *String Theory and the Scientific Method*. Cambridge: Cambridge University Press.

de Haro, S. (2017). Dualities and emergent gravity: Gauge/gravity duality. *Studies in History and Philosophy of Science Part B: Studies in History and Philosophy of Modern Physics*, *59*, 109–125. doi:10.1016/j.shpsb.2015.08.004

de Haro, S., Mayerson, D. R., & Butterfield, J. N. (2016). Conceptual aspects of Gauge/Gravity duality. *Foundations of Physics*, *46*, 1381–1425. doi:10.1007/s10701-016-0037-4

Di Casola, E., Liberati, S., & Sonego, S. (2015). Nonequivalence of equivalence principles. *American Journal of Physics*, *83(1)*, 39–46. doi:10.1119/1.4895342

Doboszewski, J., & Linnemann, N. (2018). How not to establish the non-renormalizability of gravity. *Foundations of Physics*, *48*, 237–252. doi:10.1007/s10701-017-0136-x

Donoghue, J. (1994). General relativity as an effective field theory. The leading quantum corrections. *Physical Review D*, *59*, 3874–3888.

Doplicher, S., Fredenhagen, K., & Roberts, J. E. (1995). The quantum structure of spacetime at the Planck scale and quantum fields. *Communications in Mathematical Physics*, *172*, 187–220. doi:10.1007/BF02104515

Dowker, F. (2013). Introduction to causal sets and their phenomenology. *General Relativity and Gravitation*, *45*, 1651–1667. doi:10.1007/s10714-013-1569-y

Dowker, F., Henson, J., & Sorkin, R. D. (2004). Quantum gravity phenomenology, Lorentz invariance and discreteness. *Modern Physics Letters A*, *19(24)*, 1829–1840. doi:10.1142/S0217732304015026

Duff, M. J. (1974). On the significance of perihelion shift calculations. *General Relativity and Gravitation*, *5*, 441–452. doi:10.1007/BF00763038

Duff, M. J. (1975). Covariant Quantization. In C. J. Isham, R. Penrose, & D. W. Sciama (Eds.), *Quantum Gravity. An Oxford Symposium* (pp. 78–135). Oxford: Oxford University Press.

Earman, J. (1989). *World Enough and Space-Time*. Cambridge, MA: MIT Press.

Earman, J. (2002). Thoroughly modern McTaggart: Or, what McTaggart would have said if he had read the general theory of relativity. *Philosophers' Imprint*, *2(3)*, 1–28. doi:2027/spo.3521354.0002.003

Einstein, A. (1914). Principles of Theoretical Physics. Inaugural address before the Prussian Academy of Sciences. First published in the Proceedings of the Prussian Academy of Sciences, 1914. Reprint in *Ideas and Opinions*, based on C. Seelig (Ed.), *Mein Weltbild*, and other sources. New translations and revisions by S. Bargmann, pp. 220–223. New York, NY: Crown Publishers, 1954.

Eppley, K., & Hannah, E. (1977). The necessity of quantizing the gravitational field. *Foundations of Physics*, 7, 51–68. doi:10.1007/BF00715241

Esfeld, M. (2021). Against the disappearance of spacetime in quantum gravity. *Synthese*, 199(Suppl 2), S355–S369. doi:10.1007/s11229-019-02168-y

Falls, K., & Litim, D. F. (2014). Black hole thermodynamics under the microscope. *Physical Review D*, 89(8), 084002. doi:10.1103/PhysRevD.89.084002

Feynman, R., Morinigo, F. B., Wagner, W. G., & Hatfield, B. (1995). *Feynman Lectures on Gravitation*. Reading, MA: Addison-Wesley.

Friederich, S. (2018). The asymptotic safety scenario for quantum gravity – An appraisal. *Studies in History and Philosophy of Science Part B: Studies in History and Philosophy of Modern Physics*, 63, 65–73. doi:10.1016/j.shpsb.2017.12.001

Georgi, H. (1993). Effective field theory. *Annual Review of Nuclear and Particle Science*, 43(1), 209–252. doi:10.1146/annurev.ns.43.120193.001233

Ghins, M., & Budden, T. (2001). The principle of equivalence. *Studies in History and Philosophy of Science Part B: Studies in History and Philosophy of Modern Physics*, 32(1), 33–51. doi:10.1016/S1355-2198(00)00038-1

Haack, S. (1993). Evidence and Inquiry. *Towards Reconstruction in Epistemology*. Oxford/Cambridge, MA: Blackwell.

Hatfield, B. (1995). Quantum Gravity. In R. P. Feynman, F. B. Morinigo, W. G. Wagner, & B. Hatfield (Eds.), *Feynman Lectures on Gravitation* (pp. xxxi–xl). Reading, MA: Addison-Wesley.

Hawking, S. (1975). Particle creation by black holes. *Communications in Mathematical Physics*, 43(3), 199–220. doi:10.1007/BF02345020

Healey, R. (2002). Can Physics Coherently Deny the Reality of Time? In C. Callender (Ed.), *Time, Reality and Experience* (pp. 293–316). Cambridge: Cambridge University Press.

Healey, R. (2004). Change without change, and how to observe it in general relativity. *Synthese*, 141(3), 381–415. doi:10.1023/B:SYNT.0000045127.04908.1a

Hossenfelder, S. (2013). Minimal length scale scenarios for quantum gravity. *Living Reviews in Relativity*, 16(2). doi:10.12942/lrr-2013-2

Huggett, N. (2017). Target space ≠ space. *Studies in History and Philosophy of Science Part B: Studies in History and Philosophy of Modern Physics*, 59, 81–88.

Huggett, N., & Callender, C. (2001a). Introduction. In N. Huggett, & C. Callender (Eds.), *Physics Meets Philosophy at the Planck Scale. Contemporary Theories in Quantum Gravity* (pp. 1–33). Cambridge: Cambridge University Press.

Huggett, N., & Callender, C. (2001b). Why quantize gravity (or any other field for that matter)? *Philosophy of Science*, 68(S3), S382–S394. doi:10.1086/392923

Huggett, N., Linnemann, N., & Schneider, M. D. (2023). *Quantum Gravity in a Laboratory?* Cambridge: Cambridge University Press.

Huggett, N., & Vistarini, T. (2015). Deriving general relativity from string theory. *Philosophy of Science*, 82(5), 1163–1174. doi:10.1086/683448

Huggett, N., Vistarini, T., & Wüthrich, C. (2013). Time in Quantum Gravity. In H. Dyke, & A. Bardon (Eds.), *A Companion to the Philosophy of Time* (pp. 242–261). Chichester: Wiley-Blackwell.

Huggett, N., & Wüthrich, C. (2013). Emergent spacetime and empirical (in) coherence. *Studies in History and Philosophy of Science Part B: Studies in History and Philosophy of Modern Physics, 44(3),* 276–285. doi:10.1016/j.shpsb.2012.11.003

Huggett, N., & Wüthrich, C. (book manuscript). *Out of Nowhere.*

Huggins, E. R. (1962). *Quantum mechanics of the interaction of gravity with electrons: theory of a spin-two field coupled to energy.* Ph.D. dissertation, California Institute of Technology. doi:10.7907/05S4-6910

Jaksland, R. (2021). Entanglement as the world-making relation: Distance from entanglement. *Synthese, 198,* 9661–9693. doi:10.1007/s11229-020-02671-7

Jaksland, R., & Salimkhani, K. (2023). The many problems of spacetime emergence in quantum gravity, *The British Journal for the Philosophy of Science.*

Kiefer, C. (2006). Quantum gravity: General introduction and recent developments. *Annalen der Physik, 15(1–2),* 129–148. doi:10.1002/andp.200510175

Kiefer, C. (2007). *Quantum Gravity.* Oxford: Oxford University Press.

Kraichnan, R. H. (1956). Possibility of unequal gravitational and inertial masses. *Physical Review, 101(1),* 482–488. doi:10.1103/PhysRev.101.482

Lam, V., & Wüthrich, C. (2018). Spacetime is as spacetime does. *Studies in History and Philosophy of Science Part B: Studies in History and Philosophy of Modern Physics, 64,* 39–51. doi:10.1016/j.shpsb.2018.04.003

Le Bihan, B. (2018). Space emergence in contemporary physics: Why we do not need fundamentality, layers of reality and emergence. *Disputatio, 10(49),* 71–95. doi:10.2478/disp-2018-0004

Le Bihan, B. (2021). Spacetime emergence in quantum gravity: Functionalism and the Hard problem. *Synthese, 199*(Suppl 2), 371–393. doi:10.1007/s11229-019-02449-6

Le Bihan, B., & Linnemann, N. (2019). Have we lost spacetime on the way? Narrowing the gap between general relativity and quantum gravity. *Studies in History and Philosophy of Science Part B: Studies in History and Philosophy of Modern Physics, 65,* 112–121. doi:10.1016/j.shpsb.2018.10.010

Lehmkuhl, D. (2019). General relativity as a hybrid theory: The genesis of Einstein's work on the problem of motion. *Studies in History and Philosophy of Science Part B: Studies in History and Philosophy of Modern Physics, 67,* 176–190. doi:10.1016/j.shpsb.2017.09.006

Lehmkuhl, D. (2021). The Equivalence Principle(s). In E. Knox, & A. Wilson (Eds.), *The Routledge Companion to Philosophy of Physics* (pp. 125–144). New York, NY/London: Routledge.

Linnemann, N. (2019). *Philosophy of quantum gravity as a philosophy of discovery.* Ph.D. dissertation, Université de Genève. doi:10.13097/archive-ouverte/unige:127896

Linnemann, N. (2021). On the empirical coherence and the spatiotemporal gap problem in quantum gravity: And why functionalism does not (have to) help. *Synthese, 199*(Suppl 2), 395–412. doi:10.1007/s11229-020-02659-3

Loebbert, F. (2008). The Weinberg-Witten theorem on massless particles: An essay. *Annalen Der Physik, 17(9–10),* 803–829. doi:10.1002/andp.200810305

Loll, R. (2019). Quantum gravity from causal dynamical triangulations: A review. *Classical and Quantum Gravity, 37(1)*, 013002. doi:10.1088/1361-6382/ab57c7

Malament, D. B. (1977). The class of continuous timelike curves determines the topology of spacetime. *Journal of Mathematical Physics, 18(7)*, 1399–1404. doi:10.1063/1.523436

Malament, D. B. (2012). A Remark About the "Geodesic Principle" in General Relativity. In M. Frappier, D. H. Brown, & R. DiSalle (Eds.), *Analysis and Interpretation in the Exact Sciences. Essays in Honour of William Demopoulos* (pp. 245–252). Dordrecht: Springer.

Marletto, C., & Vedral, V. (2017). Witnessing the quantumness of a system by observing only its classical features. *NPJ Quantum Information, 3*(41), 1–4. doi:10.1038/s41534-017-0040-4

Martens, N. C. (2019). The metaphysics of emergent spacetime theories. *Philosophy Compass, 14(7)*, e12596. doi:10.1111/phc3.12596

Matarese, V. (2019). Loop quantum gravity: A new threat to Humeanism? Part I: The problem of spacetime. *Foundations of Physics, 49*, 232–259. doi:10.1007/s10701-019-00242-6

Matsubara, K. (2017). Quantum gravity and the nature of space and time. *Philosophy Compass, 12(3)*, e12405. doi:10.1111/phc3.12405

Mattingly, J. (2005). Is Quantum Gravity Necessary? In A. J. Kox, & J. Eisenstaedt (Eds.), *The Universe of General Relativity* (pp. 327–338). Basel: Birkhäuser.

Mattingly, J. (2006). Why Eppley and Hannah's thought experiment fails. *Physical Review D, 73*, 064025. doi:10.1103/PhysRevD.73.064025

Maudlin, T. (1996). On the unification of physics. *The Journal of Philosophy, 93(3)*, 129–144. doi:10.2307/2940873

Maudlin, T. (2002). Thoroughly muddled McTaggart: Or, how to abuse gauge freedom to generate metaphysical monstrosities. *Philosophers' Imprint, 2(4)*, 1–23. doi:2027/spo.3521354.0002.004

McKenzie, K. (2011). Arguing against fundamentality. *Studies in History and Philosophy of Science Part B: Studies in History and Philosophy of Modern Physics, 42(4)*, 244–255. doi:10.1016/j.shpsb.2011.09.002

McKenzie, K. (2017). Against brute fundamentalism. *Dialectica, 71*, 231–261. doi:10.1111/1746-8361.12189

McKenzie, K. (2022). *Fundamentality and Grounding*. Cambridge: Cambridge University Press.

Menon, T. (2019). On the viability of the no alternatives argument. *Studies in History and Philosophy of Science Part A, 76*, 69–75. doi:10.1016/j.shpsa.2018.10.005

Misner, C. W., Thorne, K. S., & Wheeler, J. A. (1973). *Gravitation*. Princeton, NJ: Princeton University Press.

Møller, C. (1962). The Energy-Momentum Complex in General Relativity and Related Problems. In M. A. Lichnerowicz, & M. A. Tonnelat (Eds.), *Les théories relativistes de la gravitation* (pp. 15–29). Paris: Centre National de la Recherche Scientifique.

Morrison, M. (2006). Emergence, reduction, and theoretical principles: Rethinking fundamentalism. *Philosophy of Science, 73(5)*, 876–887. doi:10.1086/518746

Muñoz, D. (2020). Grounding nonexistence. *Inquiry, 63(2)*, 209–229. doi:10.1080/0020174X.2019.1658634

Ney, A. (2021). The World in the Wave Function. A Metaphysics for Quantum Physics. Oxford: Oxford University Press.

Nicolis, A. (2011). General Relativity from Lorentz Invariance. *General Relativity from Lorentz Invariance*.

Niedermaier, M., & Reuter, M. (2006). The asymptotic safety scenario in quantum gravity. *Living Reviews in Relativity, 9*, 5. doi:10.12942/lrr-2006-5

Norton, J. D. (2008). Why constructive relativity fails. *The British Journal for the Philosophy of Science, 59(4)*, 821–834. doi:10.1093/bjps/axn046

Ortín, T. (2017). Higher order gravities and the strong equivalence principle. *Journal of High Energy Physics, 152*, 1–14. doi:10.1007/JHEP09(2017)152

Paul, L. A. (2012). Building the world from its fundamental constituents. *Philosophical Studies, 158*, 221–256. doi:10.1007/s11098-012-9885-8

Pauli, W. (1941). Relativistic field theories of elementary particles. *Reviews of Modern Physics, 13(3)*, 203–232. doi:10.1103/RevModPhys.13.203

Poisson, E., & Will, C. (2014). *Gravity. Newtonian, Post-Newtonian, Relativistic.* Cambridge: Cambridge University Press.

Read, J. (2020). Explanation, Geometry, and Conspiracy in Relativity Theory. In C. Beisbart, T. Sauer, & C. Wüthrich (Eds.), *Thinking About Space and Time: 100 Years of Applying and Interpreting General Relativity. Einstein Studies, Vol. 15* (pp. 173–205). Basel: Birkhäuser.

Read, J., Brown, H. R., & Lehmkuhl, D. (2018). Two miracles of general relativity. *Studies in History and Philosophy of Science Part B: Studies in History and Philosophy of Modern Physics, 64*, 14–25. doi:10.1016/j.shpsb.2018.03.001

Redhead, M. (1999). Quantum Field Theory and the Philosopher. In T. Y. Cao (Ed.), *Conceptual Foundations of Quantum Field Theory* (pp. 34–40). Cambridge: Cambridge University Press.

Rickles, D. (2011). A philosopher looks at string dualities. *Studies in History and Philosophy of Science Part B: Studies in History and Philosophy of Modern Physics, 42(1)*, 54–67. doi:10.1016/j.shpsb.2010.12.005

Rickles, D. (2013). AdS/CFT duality and the emergence of spacetime. *Studies in History and Philosophy of Science Part B: Studies in History and Philosophy of Modern Physics, 44(3)*, 312–320. doi:10.1016/j.shpsb.2012.06.001

Rickles, D. (2014). *A Brief History of String Theory.* Berlin/Heidelberg: Springer-Verlag.

Rickles, D. (2020). *Covered with Deep Mist. The Development of Quantum Gravity (1916-1956).* Oxford: Oxford University Press.

Robb, A. A. (1914). *A Theory of Time and Space.* Cambridge: Cambridge University Press.

Rosenfeld, L. (1963). On quantization of fields. *Nuclear Physics, 40*, 353–356. doi:10.1016/0029-5582(63)90279-7

Rovelli, C. (2011). Forget time. *Foundations of Physics, 41*, 1475–1490. doi:10.1007/s10701-011-9561-4

Salimkhani, K. (2018). Quantum Gravity: A Dogma of Unification? In A. Christian, D. Hommen, N. Retzlaff, & G. Schurz (Eds.), *Philosophy of Science: Between the Natural Sciences, the Social Sciences, and the Humanities* (pp. 23–41). Cham: Springer International Publishing. doi:10.1007/978-3-319-72577-2_2

Salimkhani, K. (2020). The dynamical approach to spin-2 gravity. *Studies in History and Philosophy of Science Part B: Studies in History and Philosophy of Modern Physics, 72,* 29–45. doi:10.1016/j.shpsb.2020.05.002

Salimkhani, K. (2022). A Dynamical Perspective on the Arrow of Time. http://philsci-archive.pitt.edu/20852/

Schiff, L. I. (1960). On experimental tests of the general theory of relativity. *American Journal of Physics, 28(4),* 340–343. doi:10.1119/1.1935800

Shomer, A. (2007). A pedagogical explanation for the non-renormalizability of gravity. arXiv:0709.3555 [hep-th]

Sklar, L. (1977). What might be right about the causal theory of time. *Synthese, 35,* 155–171. doi:10.1007/BF00485494

Unruh, W. G., & Wald, R. M. (1989). Time and the interpretation of canonical quantum gravity. *Physical Review D, 40(8),* 2598–2614. doi:10.1103/PhysRevD.40.2598

Vistarini, T. (2019). *The Emergence of Spacetime in String Theory.* London/New York, NY: Routledge.

Wald, R. M. (1984). *General Relativity.* Chicago, IL: The University of Chicago Press.

Weinberg, S. (1964a). Derivation of gauge invariance and the equivalence principle from Lorentz invariance of the S-matrix. *Physics Letters, 9(4),* 357–359. Doi:10.1016/0031-9163(64)90396-8

Weinberg, S. (1964b). Photons and gravitons in S-matrix theory: Derivation of charge conservation and equality of gravitational and inertial mass. *Physics Review, 135(4B),* B1049–B1056. doi:10.1103/PhysRev.135.B1049

Weinberg, S. (1965a). Photons and gravitons in perturbation theory: Derivation of Maxwell's and Einstein's equations. *Physics Review, 138(4B),* B98–B1002. doi:10.1103/PhysRev.138.B988

Weinberg, S. (1965b). The Quantum Theory of Massless Particles. In S. Deser, & K. Ford (Eds.), *Lectures on Particles and Field Theory* (pp. 405–485). Englewood Cliffs, NJ: Prentice-Hall.

Weinberg, S. (1972). *Gravitation and Cosmology: Principles and Applications of the General Theory of Relativity.* New York, NY: Wiley.

Weinberg, S. (1979). Ultraviolet Divergences in Quantum Theories of Gravitation. In S. W. Hawking, & W. Israel (Eds.), *General Relativity: An Einstein Centenary Survey* (pp. 790–831). Cambridge: Cambridge University Press.

Weinberg, S. (1995). *The Quantum Theory of Fields. Volume I: Foundations.* Cambridge: Cambridge University Press.

Weinberg, S. (1999). What Is Quantum Field Theory, and What Did We Think It Was? In T. Y. Cao (Ed.), *Conceptual Foundations of Quantum Field Theory* (pp. 241–251). Cambridge: Cambridge University Press.

Weinberg, S., & Witten, E. (1980). Limits on massless particles. *Physics Letters B, 96(1–2),* 59–62. doi:10.1016/0370-2693(80)90212-9

Wheeler, J. A. (1984). Quantum Gravity: The Question of Measurement. In S. M. Christensen (Ed.), *Quantum Theory of Gravity: Essays in Honor of the 60th Birthday of Bryce S. DeWitt* (pp. 224–233). Bristol: Adam Hilger Ltd.

Wilczek, F. (2002). Scaling mount Planck III: Is that all there is? *Physics Today*, *55(8)*, 10–11. doi:10.1063/1.1510264

Will, C. (2001). *Theory and Experiment in Gravitational Physics*. Cambridge: Cambridge University Press.

Williams, P. (2019). Scientific realism made effective. *The British Journal for the Philosophy of Science*, *70*, 209–237. doi:10.1093/bjps/axx043

Wüthrich, C. (2005). To quantize or not to quantize. Fact and folklore in quantum gravity. *Philosophy of Science*, *72*, 777–788. doi:10.1086/508946

Wüthrich, C. (2012a). In Search of Lost Spacetime: Philosophical Issues Arising in Quantum Gravity. English Version of 'A la recherche de l'espacetemps perdu: questions philosophiques concernant la gravité quantique. In S. Le Bihan (Ed.), *La Philosophie de la Physique: D'aujourd'hui à demain* (pp. 222–241). Paris: Vuibert., 2013. arXiv:1207.1489 [physics.hist-ph]

Wüthrich, C. (2012b). The structure of causal sets. *Journal for General Philosophy of Science*, *43*, 223–241. doi:10.1007/s10838-012-9205-1

Wüthrich, C. (2018). The Emergence of Space and Time. In S. Gibb, R. F. Hendry, & T. Lancaster (Eds.), *The Routledge Handbook of Emergence* (pp. 315–326). London/New York, NY: Routledge.

Wüthrich, C., & Callender, C. (2017). What becomes of a causal set? *The British Journal for the Philosophy of Science*, *68*(3), 907–925. doi:10.1093/bjps/axv040

6 Unification

Put roughly, unification is about the realisation that seemingly unrelated phenomena, entities or theories are, in fact, related in a sense that is relevant for theory building. Unification often results in better explanations, more compact theoretical descriptions, larger theoretical scope, or more coherent and parsimonious ontological commitments. Relatedly, unification provides crucial information on issues of fundamentality: asking whether certain entities are fundamental means asking about the ontological relations of those entities to all other entities. Hence, unificatory perspectives are naturally more interesting than patchwork perspectives (cf. Cartwright, 1999). Attempts to unify test ontological dependence relations.

Therefore, assessing the potential for unification seems a good candidate as a criterion for constraining our metaphysical inferences, if we find ourselves in a situation of empirical underdetermination. The crucial point is that it is not (just) a philosophically motivated choice (namely, from questions of fundamentality) to bring in unification as a criterion. Rather, or so I argue in this chapter, unification is suggested as a criterion by physics practise itself – notably, even if we understand physics as merely aiming at explanations that are empirically adequate and theoretically consistent. Such practise of physics still obtains unification as a key *result*.

More specifically, I argue that unification is a *by-product* of physical research that is driven by the *basic methodological strategies* of physics alone – without appeal to metaphysical or metatheoretical presuppositions. This is what I call an *internal* (or methodological) explanation of why there is unification in physics. To support my claims, I investigate the actual practise undertaken in physics in paradigmatic examples of unification in Section 6.1.4. Subsequently, I spell out how this internal conception of unification can be used for constraining metaphysical inferences in Section 6.2 and conclude by reviewing the issue of the fundamentality of spacetime in light of my findings in Section 6.3.

DOI: 10.4324/9781003404149-6

6.1   Explaining Unification in Physics[1]

In the following, I call into question two widespread beliefs about unification in physics: (1) unification is an aim (or *the* aim) of physics, and (2) unification is driven by metaphysical or metatheoretical presuppositions. In contrast, I argue that (a) unification is a *result* – or, for dialectical effect: a *by-product* – of physical research and (b) unification is driven solely by physics' *basic methodological strategies*. In particular, I maintain that unification is not driven by any appeal to metaphysical or metatheoretical presuppositions. Accordingly, I need to answer the following questions: if physics does not explicitly aim at unification, then what explains the many obvious cases of unification in physics, i.e., why is there a unificatory practise in physics (or more precisely: why are there certain unificatory research programmes); and, what is the methodological "toolbox" that brings about unification, if metaphysical or metatheoretical presuppositions are not responsible?

6.1.1   *Internal and External Explanations*

Formally speaking, there are two ways to answer the first question: either we explain unification in physics *internally* or *externally*. An explanation is internal if it is only based on the basic (or genuine) aims and methods of physics – in short: an internal explanation is based on physical reasoning. On the other hand, an explanation is external, if it is based on additional presuppositions, aims, and methods, which cannot be regarded as an essential part of physics (e.g., metaphysical presuppositions) – in short: an external explanation is based on general philosophical reasoning.

What I call the basic (or genuine) methodology of physics is associated with (or conducive for) a minimal set of aims that physics necessarily sets for itself and that defines the discipline. Certainly, physics is a sufficiently directed enterprise to be expected to have such a set of aims, but admittedly there is no consensus on what aims to include.[2] Still, drawing on physics practise, I propose to take the following two as the most parsimonious, sufficiently accepted, and central aims that need to be accounted for in one way or another: *empirical adequacy* and *theoretical consistency*. What we expect from physical theories is, first and foremost, to be empirically adequate, i.e., to correctly represent the phenomena and empirical data. Second, physical theories are expected to be sufficiently consistent, i.e., to not include contradictory statements – at least not within a particular theory.

Empirical adequacy and theoretical consistency can be accounted for most directly if they are accepted as genuine aims of physics. To additionally accept unification as an aim is a stronger and therefore more controversial claim. Alternatively, one could argue that unification (as the sole aim) *implies* either empirical adequacy, or theoretical consistency, or both. But these too are stronger statements that are more costly than accepting the minimal set proposed. Without further argumentation, it is unclear why unification should automatically imply empirical adequacy. Just consider the example of grand unified theories, which are highly unified but not empirically adequate. Also, unification demands more than mere consistency, i.e., absence of contradictions; unification at least demands a very special kind of consistency. Thus, assuming empirical adequacy and theoretical consistency as the set of aims is in any case the weaker and more minimal assumption.[3]

This being said, it is important to emphasise that I am not engaging in the project of arguing in favour of some specific set of aims here. I do agree that this issue is rather involved. In particular, it may be possible to do without some versions of theoretical consistency[4] or expect theoretical consistency to be reducible to empirical adequacy.[5] Ultimately, it is not so much the aims, but the concrete methodology that matters. What I am after is that for explaining unification in physics it is sufficient to appeal to a "down-to-earth" conception of physics that gets rid of as many controversial aims, background assumptions (e.g., some elaborated theory of explanation), and methods as possible: already a deflationary understanding of physics' (internal) aims and methodological strategies is able to explain its unificatory practise.

In summary, the internal view – which I argue for – holds that those methodological strategies that are an essential part of doing everyday research in physics (e.g., inductive generalisations or resolving theoretical inconsistencies) are also responsible for physics' unificatory practise. The internal view maintains that already a minimal understanding of physics' aims and methods is headed towards unification; unification is part of the results of everyday research in physics. Against this, the external view holds either that there is a unificatory practise in physics because unification is what physics explicitly aims for, or that attempts at unification are based on metaphysical or metatheoretical presuppositions (i.e., philosophical reasoning). Hence, for the external view, unification is the *imposed* result of additionally constraining research by assumptions that are not part of physics proper.

To argue for the external view, one would have to show that the unificatory practise in physics hinges on certain external (e.g., metaphysical or metatheoretical) presuppositions or an explicit, methodologically active aim of unification. In particular, one would have to demonstrate where and how physicists do (and need to) employ external reasoning to justify research programmes (at least with respect to their pursuit-worthiness) and to methodologically account for unification. On the contrary, to argue for the internal view, one has to show that already essential methodological strategies of physics produce a more unified picture *as a by-product*. In short, explaining why and specifying in which sense physics is about unification amounts to investigating its actual practise. Accordingly, I study a few paradigmatic examples of unification in Section 6.1.4.

Now, philosophical debates on unification usually focus on other aspects. Hence, my main questions are not yet sufficiently addressed nor satisfyingly answered in the standard literature on unification. Nevertheless, here is what I take as typical philosophical explanations of the unificatory practise in physics: (1) claiming that physics generally aims at unification[6] or other epistemic values like explanation (with explanation either *being* unification,[7] or being supported by unification[8]), or (2) claiming that we can identify specific metaphysical (e.g., "unity of nature") or metatheoretical (e.g., simplicity or "beauty" of the mathematical representation) presuppositions that guide and drive scientific research towards unification.[9]

How should these explanations be classified with respect to the internal–external distinction? First, proponents of the view that unification is an aim of physics must argue either that the genuine methodology of physics associated with the minimal set of aims is sufficient to achieve this additional aim or that it is not; this ambiguity is usually not acknowledged in the literature and stems from the fact that such positions typically remain silent on how (aiming at) unification is concretely obtained methodologically. The view that the genuine methodology is sufficient is compatible with my internal view. Unification is then ultimately conceived of as a *redundant* aim: the genuine methodology, which supports the indispensable aims (empirical adequacy and theoretical consistency), also brings about unification as a by-product. Notice that this understanding arguably trivialises the claim that physics aims at unification: after all, unification is reduced to the status of being the result of other aims (as I claim) – so why should we consider it an aim at all? Why should we hold on to this ambiguous way of speaking?[10]

On the contrary, denying that the minimal set of aims is sufficient to achieve unification basically collapses into claim (2): physical research is taken to be guided and constrained by aiming at unification, or rather by certain additional methodologically active presuppositions that are associated with this aim. Without these additional constraints, the reasoning goes, physics methodology may yield empirically adequate and consistent theories, but no instances of unification. This boils down to what I reject as an external explanation of unification.

To be clear, one can, of course, redefine what I understand as physics proper such that these additional constraints are included and hence "internal". I do not take issue with alternative definitions of physics. I take issue with the specific claim that what we can observe as a unificatory practise of physics can allegedly not be explained by the minimal set of aims and the methodology associated with it. In this sense, this chapter also explicates how we should understand the "aiming at unification" talk so that it is in accordance with physics practise.

Second, arguing that metaphysical or metatheoretical presuppositions guide and drive scientific research towards unification explicitly adopts external reasoning: physics is taken to pursue a specifically enriched methodology that is biased towards a striving for unification by including metaphysical or metatheoretical presuppositions (call this a "*unificationist methodology*"). On this view, unification is driven by presuppositions that are not part of what I referred to as "physics proper". Such unificationist presuppositions include, for example, metaphysical presupposition like "there is unity in nature" or "everything is quantum", or metatheoretical presupposition like "ultimately, all of physics needs to be described by a single theory". Arguing that it is such additional presuppositions that bring about unification in physics implies that there is (or could be) a less unificatorily biased way of doing physics (i.e., the basic methodology without such additional presuppositions) which would not have a unificatory practise and could not account for unification. I shall argue that this is a misleading characterisation of how physics works.

6.1.2   The Physicist's Tale

One has to admit, however, that at first sight external explanations do seem to have a point. First, there is a long history of both successful *and* failed attempts at unification. Second, many physicists themselves subscribe to the view that finding a unified theory, which encompasses all physical phenomena, is what physics ultimately aims at; specific attempts at unification may fail, but nevertheless, it is the way to go. Third, unification talk is typically not merely descriptive but seeks to define what physics is about and how physics should proceed (in the future) on the basis of what seems to presume a specific sense of unity in nature. As Maudlin (1996) puts it:

> Today, anyone inquiring … into the current status of fundamental physical theory is virtually guaranteed to be told the following tale. In the first part of this century, physicists had verified the existence of four basic physical forces: electromagnetism, gravity, the strong nuclear force, and the weak nuclear force. Passably accurate theories of these forces individually have been developed, but those theories do not yet demonstrate any deep connection among all of the forces. The aim of physics is now to produce theories which unify these forces, which show, ultimately, that there is at base only one fundamental force in the universe, which has come to display itself as if it were many different forces.

The first step in this program has already been taken: electro-magnetism has been unified with the weak nuclear force in the electroweak theory. The other steps, though still to be achieved, have already been named, and are to occur in a particular sequence. The electroweak force is to be unified with the strong nuclear force by a *grand unified theory* (GUT), and then, in the final step, the GUT will somehow be unified with gravity in a *theory of everything* (TOE).

This image of the future course of physical theory has become so pervasive as to rank almost as dogma.

(Maudlin, 1996, p. 129)

According to this "physicist's tale", unification exhibits the status of a methodological "commitment" that "puts rather strong constraints" on physics (Maudlin, 1996, pp. 129–130), and seems at least partly motivated by metaphysical speculation. Although also Maudlin argues that the commitment to unification "is certainly ... not so strong as to override empirical support for theories that do not follow the projected path" (Maudlin, 1996, p. 130), such a practise of physics might still raise the following worry: physics (or rather mainstream research programmes in physics) might be led astray by the "dogma of unification", which "is a powerful image that helps shape the direction of research" (Maudlin, 1996, p. 130), and eventually run into a dead end since physicists do not even consider the possibility of a disunited picture anymore – a worry resembling Wolfgang Pauli's verdict on Einstein's unified field theory: "What God put apart, no one shall unify."[11] I intend to assuage such worries by demonstrating that, in fact, unification is obtained indirectly due to essential parts of the physics methodology, rendering unification a natural consequence of doing physics.

One might also argue (without specific metaphysical presumptions) that this commitment to unification should rather be understood as a mere heuristic strategy for guiding theory development – based on the meta-inductive observation that it previously had proven worthwhile to explore directions of research that are assumed to exhibit a unifying power. Kao (2019) can be taken to present such a position. I do not argue against Kao's proposal *per se*. This is because I take her reading of unification as playing a heuristic role in physics to be sufficiently weak to be compatible with what I am trying to put forward. In fact, I very much agree with Kao on what physicists are *actually* concerned with:

indeed, it does not even seem to be the case that finding an overarching theory was the primary motivation to work in the various domains. ... it is worth noting that this is the opposite of what we might think would be the case if the goal were, for instance, to find unifying theories because unifying explanations are a crucial characteristic of theories. Instead, it seems that the primary motivation was to solve existing problems in each scientific domain.

(Kao, 2019, p. 3271)

My project is that of taking this insight seriously.[12]

It strikes me that both Maudlin's physicist's tale and Kao's heuristical reading miss or disregard important aspects of how theory development in physics is obtained methodologically. As a result, such accounts of physics' practise of unification pave the way for dismissive criticism of this very practise, as for example in Mattingly (2005) or Hossenfelder (2018). In short, I take it that there is more to learn about unification in physics.

As a science, modern physics subsumes the diversity of singular events under general laws of nature. To allow for this, physics arguably needs to employ some fundamental assumptions about its subject matter, but unification (of these laws of nature), so I claim, does not depend on any additional metaphysical or metatheoretical presuppositions that are not already implicit in genuine strategies of physics themselves. It is not the case that we forcefully try to unify what God put apart. Rather, *whenever we try to understand a limited domain, we are pushed beyond its boundaries*. In the course of physical research, the concrete objects of research that physicists are concerned with contingently prove to exhibit surprisingly substantial links to other objects of research *as a matter of fact*. So, we may say that the objects of research do "effectively" *turn out* to fit together in a manner rather similar to jigsaw pieces. But to say this does *not presuppose* unity in nature, but *infers* it (or a possibly restricted version thereof).[13] Hence, the claim is that even something as speculative as the quest for a theory of quantum gravity should be understood as the *result* of an *immanent* analysis of our theoretically most advanced and experimentally best-tested framework, not as a striving for unification *simpliciter*. Accordingly, unification is "only" a by-product of exploiting genuine scientific methods. Philosophy should not obscure this important characteristic of physics by the usual talk about the alleged striving for unification.

Again, all these claims need to be substantiated by an investigation of physics practise. But before I begin to do so, let me first provide further clarification of the concept of unification.

6.1.3 *Mapping Concepts of Unification*

Typically, unification in physics is about generalising or subsuming different laws of nature in a superordinate theory in a meaningful way.[14] For example, the laws of Galileo and Kepler are often deemed "unified" in Newton's universal law of gravitation, because Newton's law comprises and explains the laws of Galileo and Kepler. But what are the exact criteria for considering some combination of laws an instance of unification?

The mere consistency of a set of laws is not sufficient to regard them as unified,[15] nor is the fact that they share a common explanatory structure or have a nomic connection, as Maudlin (1996) argues: a common explanatory structure that does not put any constraint on the explanantia does not mean that the laws are unified (Maudlin, 1996, p. 131). A theory that combines Newton's law of gravitation and Coulomb's law of electricity, for example, offers a common explanatory structure for electricity and gravitation, but since "the forces are not postulated to have anything in particular to do with one another" this is not a unification of gravity and electricity (Maudlin, 1996, p. 131). Similarly, Maudlin argues that electricity and magnetism are not unified in Maxwell's theory of electromagnetism, because their intimate nomic connection is not registered in the ontological commitments of Maxwell's theory. For Maudlin, the notion of unification requires some form of ontological import: unification is only obtained when previously ontologically independent entities or structures are reduced to a single entity or structure in the new theory. Therefore, electromagnetism is only truly unified in Einstein's theory of special relativity, where the electric field and the magnetic field are "replaced by" (Maudlin, 1996, p. 133) an underlying structure, namely the electromagnetic field.

Overall, I agree with the analysis that Maudlin provides.[16] I do, however, focus on a different issue, namely investigating the *methodological* aspects in the *process* of unification, i.e., the process of "climbing" Maudlin's conceptual ladder (starting from the diversity of the phenomena and ending – maybe – in ontological unification). Due to my focusing on the process rather than the end result, my notion of unification becomes more permissive than Maudlin's. In particular, I propose to map different types of unification according to two questions: (1) What is the object of unification? and (2) How does unification work methodologically?

The first question is easy to answer. There are three levels (or dimensions) of unification. (1.1) Unification can be about unifying a variety of phenomena by some (phenomenological) law of nature. For example, Kepler's laws unify certain aspects of planetary motion. Call this type of unification *phenomenological* (or *nomological*) unification. Furthermore, (1.2) unification can be about unifying (already existing) theoretical structure, like laws and theories, by new laws and theories. For example, Newton's theory unifies the laws of Galileo and Kepler. Call this type *theoretical* unification. Finally, (1.3) unification can be about unifying ontic entities or structures. For example, the electromagnetic field unifies electric and magnetic fields. Call this type *ontological* unification. Note that while unifying ontology generally requires theoretical and phenomenological unification, the opposite is not true. So, the strength of the unification obtained is understood to increase from (1.1) to (1.3).

Concerning the second question on how unification is actually worked out, I propose to distinguish five methodological strategies that physics regularly employs and that carry unificatory impact (as further discussed in the case studies in the next section): (2.1) inductive *generalisation* from phenomena to (phenomenological) laws of nature[17]; (2.2) *eliminating theoretical misrepresentation* of the observable phenomena (e.g., eliminating artefacts like frame dependence); (2.3) *resolving anomalies and inconsistencies* within a theoretical framework; (2.4) identifying *(causal) mechanisms*; and (2.5) exploiting the explanatory resources of an established theory in full and expanding its scope from within (*second order generalisation*). These strategies are *internal* in the sense that they are *genuine* to physics – i.e., strategies (2.1)–(2.5) constitute an essential and basic part of everyday physics research with its minimal set of aims – and do not appeal to metaphysical or metatheoretical (i.e., external) reasoning.

Constructing (or finding) laws of nature by means of inductive generalisations is widely regarded as a genuine and basic part of physics. Inductive generalisations form the very basis of physics and are therefore in a sense the starting point of unification. Nevertheless, inductive generalisations neither directly refer to unification,[18] nor to (e.g., metaphysics-laden) considerations external to physics (see also Section 6.1.4). But what about the other strategies? Do they refer directly to unification or to considerations external to physics? Do they involve some sort of metaphysical presupposition? As I shall argue in more detail below, this is not the case. In brief, strategy (2.2) is only concerned with correcting any theoretical misrepresentation of the phenomena by evaluating, for example, whether certain properties of the phenomena (e.g., symmetry properties) are represented correctly, and ensuring that the theory does not posit different theoretical entities as representation of the same phenomenon without reason; I discuss these issues in more detail when studying electromagnetism in special relativity. Strategy (2.3) appeals to the fundamental demand that theories need to be logically (or mathematically) consistent, but this does not presuppose anything about the comprehensiveness of the theory. Strategy (2.4) is a paradigmatic method of physics that does not involve any claim regarding the unity of such mechanisms; and (2.5) is an outstanding example of an *internal* unificatory practise as it simply comprises taking an empirically established theory seriously, exploiting it in full, and pushing it to its (alleged) limits *by its own means* (until anomalies arise) – certainly, it is a genuine part of physics research to work out the details of its theories and check whether something has to give.

For further clarification, let me list two conceivable strategies that are – on their own – *not* genuine to physics, but refer to external reasoning:

given two theories, (2.6) identifying a common (mathematical) structure or reducing one to the other, and (2.7) identifying a common ontology or reducing one to the other. Employing method (2.6) can be understood as enforcing the purely formal demand of mathematical simplicity on physics that merely reorganises the laws in a "simpler" way (without additional explanatory insights). Moreover, strategy (2.7) – if not a result of internal strategies – explicitly relies on philosophical considerations and does not seem to be a genuine physical method at all (physicists are typically not concerned with ontological questions in everyday research).[19] As already mentioned, some *do* think that physics employs externally justified strategies (see my discussion of the case of quantum gravity below).

Note that these strategies are not isolated, but may overlap or work together. For example, an analysis according to (2.1) inductive generalisation or (2.2) eliminating misrepresentation can result in ontological unification. Also, I do not conceive the list of methodological strategies to be exhaustive or fixed, but as allowing for additional nuances. Furthermore, while strategy (2.1) refers exclusively to level (1.1), i.e., phenomenological unification, the other strategies might in principle account for unification at all three levels.

We can now identify the two concepts of unification that are linked to the opposing ways of explaining why there is a unificatory practise in physics. The *internal view* holds that physics' genuine methodological strategies, (2.1)–(2.5), designed to uncover, exploit, and expand the explanatory resources of a theory as far as possible, are sufficient to obtain unification at all three levels, (1.1)–(1.3), as a *by-product* of physics research. It is because of its internal unificatory impact, the reasoning goes, that we should not be surprised that genuine physics methodology on its own already implies a unificatory practise. Such a view neither draws on metaphysical or metatheoretical considerations, nor on explicitly aiming at unification. Accordingly, unification *arises in* physics.

At first sight, Maudlin (1996) seems to endorse an internal view when he argues that it is possible to identify empirical or theoretical evidence *suggesting* unification in many cases. However, Maudlin's conviction that physics is committed to unification as "the aim of physical theory" (Maudlin, 1996, pp. 129–130), his presentation of the physicist's tale, and his interpretation of certain case studies (see below) are incompatible with the internal perspective I advocate here and rather point to a position that views physics as constrained and guided externally.

On the other hand, the *external view* argues either that we pursue unification for its own sake (or as being instrumental for promoting other epistemic values, like explanation,[20] for example), or that we pursue unification based on metaphysical or metatheoretical presuppositions. At the heart of both lies a unificationist methodology that executes the external demands that drive the striving for unification; strategies (2.6) and (2.7) may serve as examples of how physics' genuine methodology could be extended in the direction of a unificationist methodology. According to the external view, unification is *imposed on* physics. Hence, unification is a result of additionally constraining research by assumptions external to physics proper. The reasoning essentially works as follows: "Due to some metaphysical or metatheoretical presupposition we know that physics should pursue unification. Therefore, we try to find (and should prefer) common theoretical and ontic structures to construct unified theories." For instance, the way Maudlin presents the physicist's tale as a quest for (1.3) a unified ontology of the fundamental interactions suggests that physics pursues an externally justified unificationist methodology by explicitly trying to (2.7) identify a common ontology. Proponents of this view emphasise that physicists are – and should be – dissatisfied with any kind of patchwork in physics (e.g., completely disconnected theories for different interactions) for the respective reasons. Critics take this practise as de facto existing but misguided, and the reasons for pursing it as physically dubious.

Against both variants of the external perspective, I claim that there is no such practise of imposition, but still unification is obtained. We should not interpret physics practise in line with the external view, because it obfuscates how theory development is actually driven and justified in physics, namely by concrete physical reasons. Opposing statements by physicists on their intentions or motivations should not be taken too seriously. Instead of evaluating what physicists say, I propose to look at what they actually do.

6.1.4 Four Case Studies from Physics

In the following, I seek to support these claims by investigating four paradigmatic case studies: Newton's universal law of gravitation, Einstein's unification of electromagnetism, electroweak theory and grand unified theories, and quantum gravity.

6.1.4.1 Newton's Universal Law of Gravitation

The historically first substantial example of unification in physics is Newton's universal law of gravitation which famously unified the terrestrial and celestial mechanics of Galileo's and Kepler's laws. Equally famously, Newton was aware of and explicit about the metaphysical foundations and consequences. So, is Newton's theory an instance of unification that was driven by metaphysical reasoning?

Reading Ducheyne (2005), it would seem so. Ducheyne argues that we can identify two types of unification in Newton's work. The first type of unification is established by identifying causal mechanisms, while the second refers to inductive generalisations. Ducheyne specifies that the first type "is based on the premise that nature maintains the same *modus operandi* as much as possible", whereas the second "is based on the premise that nature is contiguous" (Ducheyne, 2005, p. 77). Regarding the first type of unification,

> Newton infers from the observation that Kepler's second and third law hold for the primary and secondary planets that these motions are caused by inverse-square centripetal forces. ... Newton concludes that since these phenomena are of the same kind they depend on causes of the same kind ....
>
> (Ducheyne, 2005, p. 74)

Ducheyne goes on to argue that after Newton has identified gravity as the common cause for planetary motion, "a deep belief in the contiguity of nature" (Ducheyne, 2005, p. 74) fuels the subsequent development towards his universal law of gravitation.

We could understand this as follows. There are two metaphysical assumptions in Newton's work: on the one hand, nature is assumed to "maintain the same *modus operandi*", on the other hand, nature is assumed to be "contiguous". Unification is then obtained by help of these assumptions. For example, unification by means of method (2.4) – identifying causal mechanisms – is based on the metaphysical assumption that nature "maintains the same *modus operandi*", i.e., the same effects have the same cause. Unification would then depend on imposing metaphysical assumptions, just as the external view has it.

Against this, I propose to take the assumption that the same effects have the same cause as a general methodological (not a metaphysical) point about how inductive generalisation needs to be approached. If all the empirical facts about some set of phenomena suggest that they are (empirically) "of the same kind", then science needs to take this seriously. Similarly, "having the same cause" is just the best explanation for "having the same effect". If all facts speak in favour of uniformity and generalisability, it would seem dubious, if science still tried to keep things apart and told an elaborate story of why "what is *really* going on" is more complex than the phenomena reveal (see also the discussion of Einstein's dissatisfaction with Maxwell's theory of electromagnetism below). In brief, there are good internal reasons for developing a more unified description, and they do not commit us to any substantial metaphysical claim.

But even a stronger reading that accepts Newton's assumptions as metaphysical (i.e., as assumptions about the world, not about physics' methodology), does not entail that the external view is correct. This is because both assumptions do not selectively guide and drive the specific endeavour of unification (e.g., unification of two already established laws of nature), but are of concern in a more fundamental way: they are part of what allows us to engage in science (or inductive generalisation) in the first place. In this sense I agree that, for example, "same effect, same cause" is a credo paradigmatically internal to physics.[21] Some "order in the world" (Mumford & Tugby, 2013, p. 11) just has to be assumed; if one wishes, one may take this as a transcendental argument about the preconditions for doing science *per se*.[22] Obviously, this does have some impact in terms of initiating unification, but it does not imply that physics additionally employs metaphysical reasoning in order to specifically account for unification.

Let me also argue in more detail why the contiguousness assumption should count as internal to physics: I propose to read this assumption as an example of internally promoting a continuity from one theory (or phenomenological realm) to another allegedly disconnected theory (or phenomenological realm) – in the sense that a theory is taken seriously and then pushed to its limits *by its own means* as far as feasible. To demonstrate the continuity between terrestrial and celestial mechanics (which hitherto had been thought to be strictly separated realms), Newton brings up the following thought experiment in his *The System of the World* (see Figure 6.1 and Newton, 1846, pp. 512–513).[23] Suppose we are on a high mountain and drop a stone so that the stone freely falls to the ground.

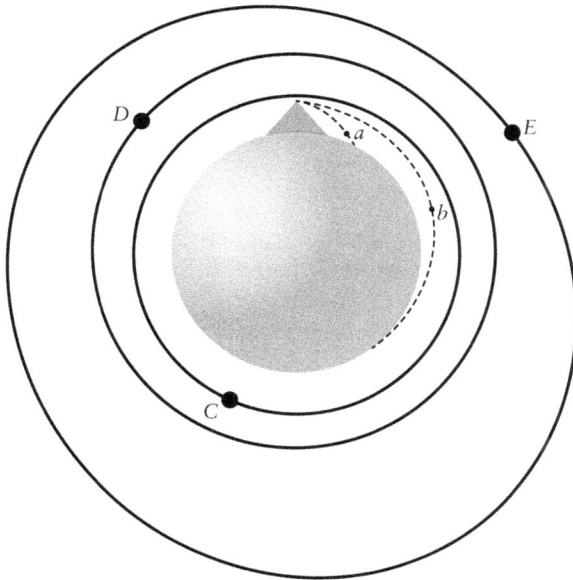

Figure 6.1 Newton's thought experiment on ballistic trajectories. Newton's thought experiment on the connection between the trajectories of stones (*a* and *b*), which are projected from a high mountain, and satellites (*C, D, E*), which orbit Earth (own recreation of Newton, 1846, p. 513).

This is described by Galileo's terrestrial mechanics. Now, suppose we do not let the stone fall, but throw it. Terrestrial mechanics tells us that the stone will move on a parabolic trajectory. When we increase the velocity (parallel to the Earth's surface) with which the stone is projected, the stone goes farther and farther. At some point the velocity is high enough to keep the stone from falling back to Earth. The stone is now orbiting Earth just as a satellite is according to Kepler's laws. By means of taking seriously the theoretical framework and realm of Galilean physics and literally pushing it to its limits (the stone's orbiting Earth is obtained as a limiting procedure of ballistic trajectories), we end up in the framework of Kepler, suggesting an underlying unified description of both; and, in particular, suggesting an underlying mechanism or common cause. Thus, Newton's unification of celestial and terrestrial mechanics is best explained by the internal unificatory impact of physics' genuine methods and can be classified as unification at all three levels (phenomena, theoretical structure, and ontology) by means of a mixture of the different internal strategies of physics.

Before we move on, let me point out that Newton's unification of the laws of Galileo and Kepler is different from the following examples in so far as it is heavily concerned with an analysis of the phenomena, whereas the following examples involve more and more prominently theoretical considerations. This demonstrates that the different examples have different points of departure regarding the already obtained level of unification – which may also explain why Newton's rules of reasoning touch on fundamental aspects of scientific methodology more directly than the following examples do.

6.1.4.2 *Electromagnetism in Special Relativity*

Maxwell's theory of electricity and magnetism is an example of what I have called phenomenological and theoretical unification. In special relativity this is pushed further: both the electric and the magnetic field are eliminated from the ontology and replaced by the electromagnetic field – which is an instance of *ontological* unification. Now, which practise of unification was at play here? Maudlin argues that Einstein identified the lack of ontological unification in Maxwell's theory as defective and explicitly aimed at unifying electricity and magnetism. In particular, Maudlin interprets Einstein's appraisal of the moving magnet and conductor problem as straightforwardly concerned with unification:

> the failure of classical electromagnetic theory to unify electric and magnetic phenomena was the leading complaint voiced in Einstein's "special relativity" paper.
>
> (Maudlin, 1996, p. 131)

This suggests that the transition from Maxwell's theory of electromagnetism to special relativity is best explained by a unificationist methodology. Yet, in fact, "the failure to unify" was *not* Einstein's complaint. Reading Maudlin's assessment, one could come away with the impression that Einstein had no physical reasons to be dissatisfied with Maxwell's theory, but preferred a unified ontology for other reasons external to physics. However, reading the passages from Einstein (1923) that Maudlin himself quotes, we find that Einstein was in fact tackling a concrete physical problem and obtained unification *as the result of a genuine physical argument*. Exploiting his principle of relativity, Einstein noticed that Maxwell's electrodynamics

> when applied to moving bodies, leads to asymmetries which do not appear to be inherent in the phenomena.
> (Einstein, 1923, p. 37; cited in Maudlin, 1996)

The same phenomena generated by the same observable current in the conductor are interpreted differently (inconsistently) in Maxwell's theory: the charges in the conductor experience either a magnetic or an electric force depending on which frame of reference is chosen for the description (the magnet frame or the conductor frame, respectively). Thus, Einstein concludes, the actual phenomenon is *misrepresented* in Maxwell's theory:

> The observable phenomenon here depends only on the relative motion of the conductor and the magnet, whereas the customary view [Maxwell's theory; my remark] draws a sharp distinction between the two cases in which either one or the other of these bodies is in motion.
> (Einstein, 1923, p. 37; cited in Maudlin, 1996)

So, Maxwell's theory artificially distinguishes two cases for one and the same *observable phenomenon* (due to frame-dependent artefacts). Einstein argues that either there is a physical explanation of the asymmetries present in Maxwell's theory (that may also be tested empirically), or the theory suffers from frame-dependent artefacts (without observable consequences) hinting at some underlying structure that is invariant under change of reference frames. Through analysing Maxwell's theory, identifying misrepresentations of phenomena, and eliminating them in special relativity, Einstein arrived at an empirically more adequate description and thereby obtained unification. Why should this be considered a case of "aiming at unification" when the point was to address a theoretical misrepresentation of a phenomenon? Just because unification was the result? For Einstein, Maxwell's theory was not defective due to a failure of unification, but due to the fact that Maxwell's theory did not consistently explain the phenomena: specifically, it did not explain why the physical situation should be different when we change the reference frame (the underlying reason for this inconsistency being that the forces present transform differently than the fields giving rise to these forces). This is important to emphasise: it is a concrete physical problem (the moving magnet and conductor problem) that raises these questions, not ideas about whether and how Newtonian and Maxwellian physics or electric and magnetic fields have to be unified. In other words, there is something to learn physically, which then promotes theoretical progress. The resulting unification was not based on any reference to external (metaphysical or metatheoretical) arguments, but was obtained by means of methods like (2.2) analysing and eliminating theoretical misrepresentation of the observable phenomena.

There are certainly more instances of unification concerning special relativity. For example, joining space and time to obtain "spacetime" is again a mere – though significant – by-product of getting the symmetry group right. But instead of investigating special relativity more thoroughly, let me now turn to a more controversial example from particle physics.

6.1.4.3 *Electroweak Theory and Grand Unified Theories*

As the physicist's tale illustrates, many are convinced that physics, amongst other things, tries to unify the three Standard Model interactions (electromagnetic, weak, and strong). Sheldon Glashow, Abdus Salam, and Steven Weinberg carried out the first step by presenting a unified theory for electromagnetism and the weak interaction: electroweak theory. However, electroweak theory is typically not interpreted as an example of ontological unification, and constructing the theory seems to be an issue of increasing formal simplicity without explanatory benefit (Morrison, 2000). Furthermore, all attempts at grand unified theories, which encompass all three Standard Model interactions, have failed. Naturally, the suspicion arises that these theories are not pursued for internal reasons of physics. Below, I counter this suspicion by highlighting that electroweak theory was the (surprising) result of the attempt to construct a consistent and empirically adequate theory that encompasses weak interaction phenomena: it only then turned out that the electromagnetic interaction is automatically included (Maudlin, 1996, p. 138). But let me approach the matter step by step. First of all, what does "unifying interactions of the Standard Model" mean? Let me briefly recall how these interactions are represented in theory.

The quantum field theories of the Standard Model are so-called gauge theories. In short, the Standard Model interactions are represented by gauge fields that arise according to the group generators of some underlying symmetry group. So, a gauge theory is essentially defined by its symmetry group. Unifying interactions then amounts to group-theoretic operations. Basically, there are three different types. First, two (or more) symmetry groups can be "glued" together to form a so-called *product group*. The product group is a trivial mathematical combination of different groups, which can always be constructed. The Standard Model itself is put together like this: the symmetry groups of the electromagnetic, $U(1)$, the weak, $SU(2)$, and the strong interaction, $SU(3)$, are merged by a trivial mathematical operation: $SU(3) \times SU(2) \times U(1)$. In Maudlin's terms "this is a case of common dynamics and nothing more" (Maudlin, 1996, p. 137). It "does not constitute any sort of unification of the theories at all" (Maudlin, 1996, p. 139). We do not learn anything above and beyond what we already knew. However, second, such a product group can, in fact, reveal additional structure. Consider, for example, electroweak theory. The unified gauge group is just the product group $SU(2) \times U(1)$, but, when combined, a non-trivial mixing between the generators of the symmetry groups occurs, such that the combination is less superficial. Third, gauge theories can be unified by absorbing their symmetry groups in a so-called *simple group*. This is not a trivial mathematical operation. One cannot simply join some groups to form a simple group. At least regarding mathematical structure, unification via some simple group is more substantial than via a product group. Physically, there are also differences. For example, an adequate choice for a simple group unifying the Standard Model usually involves additional physically relevant structure. So, in principle, one could have empirical reasons to pursue such proposals. However, consider $SU(5) \supset SU(3) \times SU(2) \times U(1)$ as an approach to grand unified theories. All such attempts at grand unified theories have failed due to the lack of empirical adequacy. Most prominently, grand unified theories predict a half-life for proton decay that is empirically excluded.

So, what drives these attempts at unification? Is it unification for the sake of formal simplicity? For electroweak theory we find this not to be the case. Rather, electroweak theory was not intended to be an attempt at unification at all. While we can write down a theory of electromagnetism alone, as obtained with quantum electrodynamics, we can*not* write down a theory of the weak interaction only[24]:

No workable theory of the weak force existed before the unified theory .... (Incidentally, this is one way in which the usual story about unifying forces is wrong. It is not that at some point we had *theories* of the electromagnetic, weak, strong, and gravitational forces separately, and now we have managed to unify the first two. Rather, at some point we *recognized the existence* of all four forces, and found that unification was needed to account for the weak force.)

(Maudlin, 1996, p. 141)

So, electroweak theory was obtained as a by-product of constructing an empirically adequate theory of the weak interaction, and hence proves to be an example of physics' internal unificatory practise; notably, despite the fact that it is typically argued that "the unification is partial, at best" (Georgi, 1989b, p. 437; in Maudlin, 1996, p. 138): electroweak theory neither uncovers a causal mechanism, nor provides ontological unification, nor exhibits any particular explanatory power beyond merely accounting for the weak force, but solely unifies mathematical structure (see Georgi, 1989b; in Maudlin, 1996), and Morrison, 2000). In my scheme, electroweak theory is therefore classified as a unification at the level of (1.1) phenomena and (1.2) theoretical structure by means of different internal strategies of physics (some mixture of strategies (2.1)–(2.5)).

What about grand unified theories? Here, the problem seems to be different. We already have a worked-out theory for the strong interaction: quantum chromodynamics. Why unify two already existing theories? A closer look reveals that grand unified theories were considered, because physicists expected that, due to the mathematical structure of these theories, grand unified theories could provide additional information with higher explanatory power for electroweak theory:

> The motivation for the simplest GUT [grand unified theory], SU(5), was not any mystical desire to follow in Einstein's footsteps and unify everything. Shelley Glashow and I were just trying to understand SU(2) × U(1) better. For several years, we had realized that if we could incorporate the SU(2) × U(1) gauge symmetry into a single simple group it would give us some extra information. It would fix the value of the weak mixing angle, a free parameter in the SU(2) × U(1) theory and it would explain why all the electric charges we see in the world are multiples of the charge of the electron.
> (Georgi, 1989a, p. 454; in Maudlin, 1996, p. 142)

Generally, such models are not analysed out of "mystical desires" (Georgi, 1989a, p. 454) for ontological unity or for the sake of formal simplicity, but to better understand the explanatory features and problems of an already established theory – by means of strategies (2.3), (2.4), and (2.5). Yet, Maudlin summarises Georgi's comment – similar to how he read Einstein's comment on Maxwell's theory – by invoking what Georgi seems to have explicitly rejected: "these considerations all stem from the desire to complete the only partially unified SU(2) × U(1) theory in a more satisfactory way" (Maudlin, 1996, p. 142). This underlines that Maudlin is interested in something other than distinguishing whether unification is driven by internal or external reasoning.[25]

6.1.4.4 *Quantum Gravity*

Such talk about physicists aiming at unification for (allegedly) physically dubious reasons is especially common and problematic when it comes to attempts at the frontier of physical research. Here, we often do not have much data to back new theoretical developments. Take quantum gravity, for example, which is usually presented as the quest to unify general relativity and quantum mechanics (or quantum field theory). When it comes to quantum gravity – the most prominent and most controversial case of unification in modern physics – we do not have any clear-cut, novel experimental or observational data at all. Additionally, the research programme of quantum gravity faces serious theoretical problems. According to folklore, this is because general relativity and quantum field theory seem to exhibit a "fundamental incompatibility" (Maudlin, 1996, p. 143). So, why unify?

As presented in Section 5.1.2, most physicists and philosophers argue that there are many physical reasons to expect an underlying theory due to phenomena where both, general relativity and quantum field theory, are understood to become relevant. Again, amongst these physical reasons are problems within the standard model of cosmology, black holes, and consistency issues in quantum field theory (Kiefer, 2007). Overall, these issues are taken to indicate some sort of incompleteness of present-day physics and trigger theoretical progress. Accordingly, the reason to pursue a theory of quantum gravity is to solve these genuinely physical issues – evidently an internal enterprise.

Recall, however, that some critics disagree and argue that there is in fact no compelling physical reason at all to pursue these research programmes – neither experimental, nor theoretical. Naturally, they advocate the disunitist's position of semi-classical theories (e.g., Mattingly, 2005). In the past, or so they argue, physics may have profited from (what they now consider a dubious) striving for unification, but modern research programmes like quantum gravity are misguided in their (alleged) commitment to unification. So have we finally encountered an example where unification is indeed driven by "purely aesthetic reasons" (Maudlin, 1996, p. 141), or "on the general methodological grounds of repeating strategies that succeeded in the past" (Maudlin, 1996, p. 141)?[26]

In fact, also philosophers and physicists who generally accept the need for a theory of quantum gravity frequently mention "aiming at unification" or a "desire for unification" as an additional reason (e.g., Crowther, 2021; Kiefer, 2007; Wüthrich, 2005), and raise similar concerns – at least regarding the motivation to pursue specific approaches, for example in the context of particle physics (Hossenfelder, 2018). For instance, Wüthrich argues that

> A strong, but nevertheless often nebulous, desire to present a unified theoretical framework at the level of fundamental physics populates

the folklore of physicists and often fuels the search for a quantum theory of gravity. Arguments to this effect, relying – if made explicit at all – on metaphysical considerations, typically elicit some principles of unity of nature or of scientific method. Although general relativity and quantum theory may be so disparate as to disallow the formal deduction of contradictions, they are generally taken to be incommensurable (families of) theories. A quantum theory of gravity is expected to remedy this theoretical schism and to bolster attempts at finding the Holy Grail of physics, a unified framework of all interactions. The argument from unification – unification for the sake of unification – does not, however, sway the sceptic. The "disunitist" would certainly be free to respond that at the very least, it may just as well be the case that the conceptual disunity of the two theories reflects a disunity in nature. In fact, she could claim, gravity's stubborn refusal to be subsumed under the otherwise all-encompassing umbrella of the Standard Model may be interpreted as evidence for this disunity. Despite its rare explicit articulation and its questionable metaphysical strength, however, the unificatory impetus provides an extremely important motivation for attempts at quantizing gravity.

(Wüthrich, 2005, p. 788)

Pushing Wüthrich's worry further, Mattingly (2005) is convinced that "implicit as well as explicit philosophical motivations" (Mattingly, 2005, p. 326) and "meta-theoretical commitments of some kind" (Mattingly, 2005, p. 326) drive the research programme.[27] In order to support his claim, Mattingly refers to the following statement by Carlo Rovelli:

We have learned from GR [general relativity] that spacetime is a dynamical field among others, obeying dynamical equations, and having independent degrees of freedom: a gravitational wave is extremely similar to an electromagnetic wave. We have learned from QM [quantum mechanics] that every dynamical object has quantum properties, which can be captured by appropriately formulating its dynamical theory within the general scheme of QM.

Therefore, spacetime itself must exhibit quantum properties. Its properties ... including ... metrical properties ... must be represented in quantum mechanical terms. Notice that the strength of this "therefore" derives from the confidence we have in the two theories, QM and GR.

(Rovelli, 2001, p. 109)

Does this reveal that Rovelli uses "some kind of thesis about the unity of nature" (Mattingly, 2005, p. 332) and that "an important meta-theoretical impetus for quantising gravity follows from notions of unification" (Mattingly, 2005, p. 332), as Mattingly argues? Rovelli's verdict regarding the "extreme similarity" and its relevance arguably does seem to express some metaphysical presupposition, but had better be read as being about comparing the theoretical concepts of two theories, and *then* drawing conclusions about the world. Accordingly, some (metaphysically more modest) story in line with Kao (2019) better accounts for what is going on here. Still, Rovelli's statement on the lessons from quantum mechanics begs the question, because the level of generality – "every dynamical object" (Rovelli, 2001, p. 109) – is chosen with respect to the desired (rhetorical) argument in favour of unification. So, on the face of it, the critics seem to have a point.

However, as emphasised before, philosophy should not get distracted by programmatic remarks. What is crucial is an evaluation of the actual practise of physics. Due to the multitude of different approaches this cannot be done exhaustively for all quantum gravity research programmes here. Instead, I will again focus on the particular class of approaches in the context of particle physics and string theory that is – in line with the physicist's tale – often thought to most clearly exemplify a striving for unification. Alternative approaches to quantum gravity usually have a rather different and less comprehensive objective. Loop quantum gravity, for example, is merely concerned with a quantum theory of (empty) spacetime. Notably, all approaches *concretely* engage with what they take to be key physical insights of either quantum field theory or general relativity. This suggests that also these approaches do not appeal to philosophical reasoning, but try to work out solutions to specific physical problems.

In the following, I will argue that we can make Rovelli's remarks more precise and thereby rebut the critics who do not accept the mentioned physical reasons to engage in the research programme of quantum gravity. This is done by emphasising the substantive insights from quantum field theory.

As argued in Section 5.2 – summarising the standard view of particle physics on this issue based on the work of Weinberg and others – a close analysis of the framework of quantum field theory delivers an effective theory of quantum gravity that is empirically adequate, but has specific theoretical problems that need to be resolved. In particular, this approach results in a low-energy theory of quantum gravity that has general relativity as its classical limit – this *is* a working theory of quantum gravity. However, the theory becomes inconsistent in the most interesting regime, namely at high energies (the infamous renormalisation problem).[28] Not only does quantum field theory provide us with a first working theory of quantum gravity, but *also* shows that we need to go beyond. Furthermore, recall that general relativity is of no concern – we may actually assume that we did not know about general relativity. As a result, the issue of quantum gravity is not to unify two theories of allegedly separate realms (general relativity and quantum field theory), but to fix internal issues of quantum field theory, i.e., to find the correct high-energy theory.

In a first step, Weinberg's argumentation establishes a theory of quantum gravity. This is an example of physics' internal capacity to promote theoretical progress by fully exploiting an existing theoretical framework and thereby uncovering substantial links to allegedly disconnected phenomena and theoretical structures, i.e., strategy (2.5). Taking the best theoretical framework seriously, and analysing it thoroughly, results in a low-energy theory of quantum gravity. In a second step, we then find that quantum field theory does not provide a meaningful theory of gravity at high energies, again employing strategy (2.5). This is the remaining problem of quantum gravity that needs to be resolved for purely internal reasons, namely for reasons of consistency of our best theory: quantum field theory. So we pursue strategy (2.3). In fact, closely related live approaches like asymptotic safety and string theory (Dawid, 2013; Weinberg, 1999) are the result of such an analysis of quantum field theory (and may also help to better understand quantum field theory).[29]

6.1.5   Unification Revisited

Let me briefly retrace my steps. I started by observing that over time the theories of physics have become more and more unified, and that this calls for an explanation. I noticed that unification being the ultimate aim of physics (or being instrumental for other epistemic values) is what many would accept as an explanation, although it is unclear what exactly this implies methodologically. Furthermore, for some (e.g., Hossenfelder, 2018; Mattingly, 2005) the genuine methodology of physics, a tendency towards unified theories in the past, and previous success of certain heuristic strategies are not able to justify present research programmes – be it in principle or as a matter of fact. In this perspective, especially certain research programmes at the frontier which (if successful) promise unification as one of their key features look suspiciously like a misguided striving for unification – in the specific sense that physicists either naively try to repeat previous successes or start from metaphysical or metatheoretical presuppositions which turn out to severely constrain (and potentially mislead) their research. The critics' view, in short, is that unification is "imposed" on physics (for external and potentially dubious reasons).

Against this, I emphasised that a more rigorous study of the actual practise of unification in physics reveals that unification *arises* from genuine methods of physics alone. Unification does not depend on metaphysical presuppositions, but is the result of good scientific practise. Unification is best explained by a certain set of basic methods used in everyday physics. More specifically, unification is best explained by the practise of taking a concept or a theory seriously, thoroughly working out its details, pushing it to its limits, and thereby gaining new knowledge.[30] This is the "best" explanation because it explains the unificatory practise from *within* physics (without assuming additional or controversial aims and methods). Furthermore, the opposing position (1) would have to say how exactly physics methodologically incorporates the alleged external constraints (either this will involve implausible statements like "physics is directly concerned with working out a minimal ontology" or it effectively collapses into my proposal), (2) cannot provide any explanatory benefit which could justify their additional metaphysical assumptions, and (3) obfuscates the actual practise of physics.

In this respect, my project is that of reducing the share of metaphysical (or metatheoretical) presuppositions in explaining actual practise in physics. Note that my project is not that of arguing against any metaphysical basis (or formal constraints) for scientific endeavours: the position I propose does still allow for physics making fundamental assumptions about its objects, in the sense that physics may have to assume that the objects are sufficiently well-behaved to be objects of physical research in the first place. Regarding unification, however, metaphysics is not what we start with but where we end up: it is the (inductive) result of physics (depending on it and being continuous with it). So even if those methods which I take to be genuine to physics are not "metaphysically thin" but loaded with some basic (metaphysical) assumptions, we are still better off not viewing unification as driven by metaphysical or metatheoretical presuppositions. Note that it is plausible that these basic assumptions actually *do the work*, but this is still different from what Wüthrich (2005), Mattingly (2005), and others put forward.[31]

Apart from lessons in general philosophy of science, this is an interesting result for metaphysics as well, especially for naturalistically inclined programmes that employ empirical data and inductive reasoning, like inductive and naturalised metaphysics.[32] So, let me briefly comment on how arguing that unification in physics does not depend on metaphysical or metatheoretical presuppositions may be exploited further for addressing the *selection problem* that I highlighted before.

6.2   Constraining Metaphysical Inferences

Recall that I have argued in Sections 5.1.4 and 5.3 that, *prima facie*, all theories and all tenable interpretations of theories which are compatible with the available empirical data are relevant for our ontological inferences. Especially at the frontier of physical research (e.g., quantum gravity research), this defines a rather weak constraint: many theories are compatible with the scarcity of empirical data (including general relativity in its different interpretations). In particular, the ontological commitments of the relevant theories often differ so significantly that this common core constraint is hardly informative.

Apparently, only by further constraining the set of theories, namely by appeal to additional criteria, can we draw more concrete metaphysical conclusions. Apart from the specific and potentially *ad hoc* criteria that I discussed and dismissed in Section 5.1.4, it has been proposed that criteria like simplicity or unification, in an external metatheoretic sense, can be used to constrain metaphysical inferences (Paul, 2012). This is usually understood to apply "a priori reasoning" (Paul, 2012, pp. 11–12). In fact, even programmes like inductive metaphysics do not generally oppose elements of a priori reasoning, precisely for the reason to be able to further constrain metaphysical inferences. The fact that inductive metaphysics uses such methods does not immediately challenge the programme's originality. Still, such practise can be seen as taking a potentially poorly justified "loophole".

Especially the generality of these criteria, i.e., their independence of any specific physical context or justification for their application, seems suspicious. As I have argued above: if we want to take physics seriously for metaphysics, then why would we impose a specific form of ontology (or specific features thereof) as generally preferable? Why should it be the case that the most simple or unified ontology is correct *per se*?

It is here where an internal understanding of unification helps to sidestep these problems. The internal view shows that the use of unification as a metaphysically relevant criterion is not necessarily externally imposed. *In concrete cases*, it is physics itself which shows that unificatory issues play a role, although unification is not carried out in full (yet); for example, physics uncovers a robust link between supposedly isolated theories. *This* justifies constraining the set of theories accordingly. If, in a concrete case, physics indicates a unificatory perspective, i.e., if physics indicates that two theories exhibit a meaningful connection, then this is a *result of physics* that should be taken seriously for metaphysical considerations, as all results of physics should.

It is non-trivial that theories, or specific interpretations thereof, indicate connections to other parts of physics, even if the path towards unification is not worked out in full yet. Consequently, these *concrete* cases need to be investigated and assessed from within,[33] which is different from making the case that unification should be accepted as a general criterion, for example, by appeal to meta-inductive reasoning.[34]

In particular, such results allow to formulate specific *continuity conditions* for two sets of theories A and B (potentially including various interpretations), which are not (yet) ontologically unified, but exhibit a robust and sufficiently meaningful connection. The continuity conditions then select those theories and interpretations in A and B, which better meet the continuity conditions. Only the selected theories and interpretations have ontological import.

This does not impose external constraints on physics, because the continuity conditions are always concrete internal results of physics. As it stands, such evaluations do not even presuppose that physics needs to be considered as a whole, because what theories are taken into account is again only a result of physics itself: as long as physics does not establish an internal connection between two theories, there are no continuity conditions that could be considered. But as soon as this happens, and we have seen that it occasionally does, the corresponding insight must be taken seriously.

Accordingly, the context of unification explicates the sense in which a physical theory or interpretation thereof is "better" with respect to metaphysics. This should be no surprise: after all, (naturalistically inclined) metaphysics is, amongst other things, about investigating what is fundamental. This means clarifying the ontological dependence relations between the entities E_i (with $i = 1, 2, ..., n$) that are postulated by physical theories. Examining whether a candidate entity E_i is ontologically independent is essentially an inductive undertaking. In principle, one needs to check whether entity E_i is ontologically independent of entity E_j for all $i \neq j$. Whether some entity is ontologically independent has to be assessed in light of the dependence relations of that entity to *all* other entities. In this sense, fundamentality claims always concern the *whole* structure of the world. Fundamentality is at its core an inherently holistic endeavour. Therefore, considering lessons from unification suggests itself. Moreover, general external criteria, like simplicity or parsimony, are of no decisive help; the world might just not be simple or parsimonious.

It is therefore essential to consider all available dependence relations between the entities that are postulated by physics. This favours unificatory perspectives on theories and their interpretations, not because of any metaphysical desire for unity, but because such perspectives provide the most explicit information on the dependence relations; for instance, an interpretation of general relativity that is – or, rather, tries to be – self-sufficient (e.g., the standard interpretation) provides less information than an interpretation that connects to other parts of physics (e.g., the dynamical spin-2 interpretation).

In other words, unification is related to the process of testing whether entities are ontologically independent or not. This does not predetermine the potential results. It may very well be that physics is unable to provide a fully unified picture, especially in the very specific sense of the physicist's tale. It is perfectly conceivable, even likely, to assume that the unification process will stop somewhere (both in physics and in metaphysics). Then we will have obtained a set of ontologically independent entities. Whether we can see more or less directly that these entities are ontologically independent because they are part of one and the same theory, or whether we have to infer indirectly that they are independent, because they are part of *different* theories which show *no connection whatsoever*, is another matter.

Accordingly, the same reasons for rejecting that physics aims at unification apply to the case of metaphysics. Just because metaphysics arguably seeks to harmonise the posits of physical theories in the specific sense of seeking to uncover a set of ontologically independent entities, does not mean that metaphysics seeks "unity". At most, it means that metaphysics seeks consistency with respect to dependence relations.

In summary, ontological inferences on the basis of individual theories are constrained by the physical practise of unification, which reveals substantial links between supposedly isolated areas of physics.

6.3   Spacetime and Unification

With respect to the metaphysics of spacetime, the above translates to the following. The ontological inferences in the philosophy of spacetime are constrained by the fact that general relativity and the quantum field theories of fundamental particle physics are connected; most importantly, general relativity is derivable from quantum field theory. It is important that this is not a question of whether there is a similarity in the mathematical form: general relativity is derivable from the framework of quantum field theory *when explicitly considering the other quantum fields and their interactions*.

Amongst others, this gives rise to the following two continuity conditions: (1) the quantum spin-2 field couples universally and has g as its classical limit; and (2) metrical aspects cannot be fundamental in a quantum context, because the Heisenberg collapse argument tells against the fundamentality of metric structure in a quantum-theoretical framework, which brings in further support for interpreting the Minkowski metric as non-fundamental.

Other continuity conditions based on insights from general relativity add further support; for example, the fact that general relativity is an effective theory, or its treatment of matter and the explanatory role of matter field properties for chronogeometricity. Note that, in general, such continuity conditions are of different strength. Notably, the applicability of such continuity conditions to the context of spacetime philosophy has been established by concrete physical arguments, not some general appeal to unification.

As a result, interpretations of general relativity which do not square well with the continuity conditions of particle physics, should not be considered as having ontological import. In particular, spacetime fundamentalism cannot be considered a tenable ontological position: the fundamentality of g is excluded by the first continuity condition, and the fundamentality of η is excluded by the second continuity condition. But interpretations of general relativity, which *do* fit the continuity conditions (e.g., the dynamical spin-2 interpretation of general relativity), do have ontological import and constrain the extent to which spatiotemporality is non-fundamental – notably, pending further empirical results. If new empirical data suggest that quantum spin-2 theory is empirically excluded, but general relativity is not, then the metaphysical conclusions need to be revised.

To summarise, the fact that *all* theories, which are compatible with the available empirical data, have interpretations in which metrical aspects of spacetime are *non-fundamental*, whereas only some theories have interpretations in which metrical aspects of spacetime are *fundamental*, suggests the inductive metaphysical inference that metrical aspects of spacetime are indeed non-fundamental. This conclusion is not based on speculative theories of quantum gravity alone, but on general relativity as well, which arguably strengthens the result.[35] If properly interpreted, i.e., if interpreted in consideration of its connection to quantum field theory and, thereby, its connection to issues of quantum gravity, the classical theory of general relativity supports the view that metrical aspects of spacetime are non-fundamental; note again that this support from general relativity is important because there are no clear-cut empirical results that would allow to dismiss general relativity as irrelevant.

This also illustrates that the programme of spacetime constructivism is *not*, to paraphrase Nerlich (1994, p. 43), about "wielding Occam's razor for its own sake".[36] Rather, it is about finding the most coherent interpretation of physics that does justice to physics' internal trend towards unification. *Nor* is it simply a matter of giving the most coherent interpretation of (a posited) "physics as a whole": it is not what we postulate, but what physics delivers, that justifies broadening the scope of our philosophical interpretation.

Notes

1 This section is based on work previously published in Salimkhani (2021). The text has been revised and extended and is reproduced with permission from Springer Nature.
2 For example, see Hüttemann (1997).
3 Similarly, we may additionally accept that physics aims at explanations, for example, and also includes corresponding methodologies, like inferences to the best explanation. I take it that this too would make the job of explaining why

physics enjoys a practise of unification easier. Hence, I try to not explicitly rely on it. The subsequent discussion does not exclude this possibility, though.

4 See Crowther and Linnemann (2019).

5 See Steinberger (2017) with respect to general issues of rationality: "The thought is that any instance of my violating (NC) [the requirement to not both believe A and $\neg A$] is eo ipso an instance in which my beliefs are out of whack with the evidence. For when I hold contradictory beliefs, at least one of the beliefs must be unsupported by the evidence."

6 For an overview see Hüttemann (1997), for example.

7 One can draw on Friedman (1974) and Kitcher (1981), for example.

8 See Salmon (1984; 1990).

9 See, for example, Mattingly (2005) and Hossenfelder (2018). See also Cat (1998) reviewing the history of debates on unification among physicists.

10 This might be countered by pointing out that aiming for unification, albeit systematically redundant (which typically is not explicitly acknowledged), may still play an important heuristic role, as Kao (2019) argues. This is generally consistent with the internal view of unification.

11 My translation of "Was Gott getrennt hat, soll der Mensch nicht vereinen" (Treder, 1983).

12 Again, what Kao highlights here suggests a rather weak understanding of the alleged heuristic role of unification.

13 As a position which refrains from metaphysical presuppositions in explaining the tendency towards unification, the internal perspective starts neither from a unitist's (e.g., von Weizsäcker, 1980) nor from a disunitist's (e.g., Cartwright, 1999) view – insisting on either "unity" or "disunity" in nature. Still, ultimately, we may favour one view over the other by help of an inference to the best explanation (see also Section 6.1.5).

14 Rather than merely harmonising physics' language and terminology, for example.

15 "The fact that a theory of embryonic development does not contradict a theory of the formation of the rings of Saturn" does not render the two unified (Maudlin, 1996, p. 130). It may still be the case, though, that unification is derivative on a more sophisticated notion of consistency that requires, for example, a sufficiently substantive engagement between the two theories (e.g., specification of all dependence relations).

16 Similar views on unification are expressed by Morrison (2000) and Rickles and French (2006), for example.

17 Again, Kepler's laws are examples of such generalisations. Note also that this includes lessons from new empirical evidence; in the light of new observations, a generalisation may require some modification.

18 As mentioned, the standard for calling something "unification" is typically higher.

19 One might object as follows: why not say that ontological unification is a theoretical desideratum in physics? Since physical theories should be about the world, they should be about ontology. Ontological considerations would then, *pace* me, be internal to physics: often, genuine physical arguments just are arguments for ontological unification – I thank Isaac Wilhelm for raising this point. I think there are three issues here. First, I do not disagree that genuine physical arguments can result in ontological unification. I am claiming that this being the case does not imply that ontological considerations are themselves part of the genuine methodology of physics. Second, I take it that most physicists (and philosophers maintaining that physics is about empirical

adequacy, for example) would not be on board with saying that ontological considerations are part of physics. Physics certainly is about the world, but it is specifically about the empirical (observable) aspects of the world. Third, whether ontological considerations are internal to physics or not is irrelevant with respect to the question of how to explain the unificatory practise: claiming that ontological considerations are internal to physics is not sufficient to account for ontological unification. Being about ontology is not synonymous with being about a unified ontology. The real job (to answer why ontological considerations lead to a unified ontology) still remains to be done. For this we can again either appeal to external, i.e., philosophical reasons, or try to spell out internally why unification is the result of genuine physical arguments. This is precisely what I put forward from the outset: often, genuine physical arguments turn out to promote ontological unification (as a by-product).

20 I shall remain neutral on the issue of whether there is any conceptual link between unification and explanation. In general, I conceive of my position on unification as compatible with views on which physics aims at explanation or truth.

21 I thank Isaac Wilhelm for pressing me on this.

22 See, for example, Mumford and Tugby (2013).

23 Note that the "Moon-test" (Principia, Book III, Proposition IV) provides a more precise, quantitative argument based on empirical data, namely for the fact that "the force by which the moon is retained in its orbit is that very same force which we commonly call gravity; for, were gravity another force different from that, then bodies descending to the earth with the joint impulse of both forces would fall with a double velocity ...; altogether against experience" (Newton, 1846, p. 392).

24 Let me stress that, strictly speaking, this presentation adopted from Maudlin (1996) is somewhat imprecise: we did know not only some phenomenological details about the weak interaction, but also already had some effective theoretical modelling of it. Still, it is correct to say that a sensible theory of the weak interaction was not worked out yet.

25 On a side note, after attending an online lecture by Howard Georgi in December 2021, Georgi's comment has been put into perspective for me. Perhaps the best way to understand it is as "not a mystical desire to unify everything ... *but something specific*".

26 Arguments which Maudlin already dismissed in the case of grand unified theories as not convincing.

27 See also endnote 27 in Chapter 5.

28 I am not suggesting that a theory is "correct" if and only if it is UV-complete. Linnemann and Crowther (2018) nicely evaluate this issue.

29 The internal–external distinction may also be used to classify approaches as more or less promising. Considering the amount of disagreement between the different approaches to quantum gravity it should come as no surprise if certain approaches eventually become viewed as poorly justified.

30 This is not only characteristic of physics, but of mathematics as well. Consider the example of number theory where exploiting the notion of "number" further and further ("pushed to its limits") ultimately provided substantial links to other areas of mathematics. I thank Andreas Bartels for this.

31 There is another important aspect to my presentation: arguing that unification will come out as a result of proper research implies a certain sense of necessity. Within a particular line of thought (concept or theory), the tendency towards a more unified picture is inevitable – at least up to the point where the concept or

theory breaks down and is not useful anymore. This view is to be distinguished from, for example, a "best systematisation" view that can only make sense of unification in terms of a strong and explicit dependence on external considerations regarding simplicity, for example.

32 Inductive metaphysics is a recent approach to metaphysics, which arguably asserts a specific methodological continuity between science and metaphysics. It is important to stress (against certain forms of naturalised metaphysics) that inductive metaphysics takes seriously that we should not read off metaphysical commitments from a particular scientific theory alone, but also consider available alternative theories (or alternative interpretations, for that matter). For an introduction to inductive metaphysics see Engelhard et al. (2021).

33 Linnemann (2020) can be read to argue accordingly, but from a heuristic point of view.

34 With respect to defending anti-foundationalism, McKenzie (2022) has recently argued that "internal" approaches, i.e., approaches "from within the perspective of an extant physical theory" (McKenzie, 2022, p. 58) are preferable to meta-inductive approaches. Her understanding of "internal" seems more restrictive. But the overall attitude is arguably similar.

35 Typically (for a notable exception see Le Bihan and Linnemann, 2019), the arguments from quantum gravity alone are taken to indicate that spacetime is non-fundamental, *contrary* to the alleged results from general relativity.

36 According to Nerlich (1994, p. 43), "[t]he best motive for relationism seems to be that we *can* use the Razor on it" (emphasis added). Its being, in his opinion, unjustified is what he considers "a dangerous pastime" (Nerlich, 1994, p. 43).

References

Cartwright, N. (1999). *The Dappled World*. Cambridge: Cambridge University Press.

Cat, J. (1998). The physicists' debates on unification in physics at the end of the 20th century. *Historical Studies in the Physical and Biological Sciences, 28(2)*, 253–300. doi:10.2307/27757796

Crowther, K. (2021). Defining a crisis: The roles of principles in the search for a theory of quantum gravity. *Synthese, 198*(Suppl 14), 3489–3516. doi:10.1007/s11229-018-01970-4

Crowther, K., & Linnemann, N. (2019). Renormalizability, fundamentality, and a final theory: The role of UV-completion in the search for quantum gravity. *The British Journal for Philosophy of Science, 70(2)*, 377–406. doi:10.1093/bjps/axx052

Dawid, R. (2013). *String Theory and the Scientific Method*. Cambridge: Cambridge University Press.

Ducheyne, S. (2005). Newton's notion and practice of unification. *Studies in History and Philosophy of Science Part A, 36(1)*, 61–78. doi:10.1016/j.shpsa.2004.12.004

Einstein, A. (1923). On the Electrodynamics of Moving Bodies. In H. A. Lorentz, A. Einstein, H. Minkowski, & H. Weyl (Eds.), *The Principle of Relativity* (pp. 35–65). New York, NY: Dover. Reprint. Translated by W. Perrett and G. B. Jeffery. Translated from "Zur Elektrodynamik bewegter Körper", Annalen der Physik, 17, 1905.

Engelhard, K., Feldbacher-Escamilla, C. J., Gebharter, A., & Seide, A. (2021). Inductive metaphysics: Editors' introduction. *Grazer Philosophische Studien*, *98*, 1–26. doi:10.1163/18756735-00000129

Friedman, M. (1974). Explanation and scientific understanding. *The Journal of Philosophy*, *71(1)*, 5–19. doi:10.2307/2024924

Georgi, H. (1989a). Effective Quantum Field Theories. In P. Davies (Ed.), *The New Physics* (pp. 446–457). Cambridge: Cambridge University Press.

Georgi, H. (1989b). Grand Unified Theories. In P. Davies (Ed.), *The New Physics* (pp. 425–445). Cambridge: Cambridge University Press.

Hossenfelder, S. (2018). *Lost in Math: How Beauty Leads Physics Astray*. New York, NY: Basic Books.

Hüttemann, A. (1997). *Idealisierungen und das Ziel der Physik. Eine Untersuchung zum Realismus, Empirismus und Konstruktivismus in der Wissenschaftstheorie.* Berlin/New York, NY: de Gruyter.

Kao, M. (2019). Unification beyond justification: A strategy for theory development. *Synthese*, *196(8)*, 3263–3278. doi:10.1007/s11229-017-1515-8

Kiefer, C.. (2007). *Quantum Gravity*. Oxford: Oxford University Press.

Kitcher, P. (1981). Explanatory unification. *Philosophy of Science*, *48(4)*, 507–531. doi:10.1086/289019

Le Bihan, B., & Linnemann, N. (2019). Have we lost spacetime on the way? Narrowing the gap between general relativity and quantum gravity. *Studies in History and Philosophy of Science Part B: Studies in History and Philosophy of Modern Physics*, *65*, 112–121. doi:10.1016/j.shpsb.2018.10.010

Linnemann, N. (2020). Interpretations of GR as Guidelines for Theory Change. In C. Beisbart, T. Sauer, & C. Wüthrich (Eds.), *Thinking About Space and Time: 100 Years of Applying and Interpreting General Relativity. Einstein Studies, Vol. 15* (pp. 153–171). Basel: Birkhäuser.

Mattingly, J. (2005). Is Quantum Gravity Necessary? In A. J. Kox, & J. Eisenstaedt (Eds.), *The Universe of General Relativity* (pp. 327–338). Basel: Birkhäuser.

Maudlin, T. (1996). On the unification of physics. *The Journal of Philosophy*, *93(3)*, 129–144. doi:10.2307/2940873

McKenzie, K. (2022). *Fundamentality and Grounding*. Cambridge: Cambridge University Press.

Morrison, M. (2000). *Unifying Scientific Theories*. Cambridge: Cambridge University Press.

Mumford, S., & Tugby, M. (2013). What Is the Metaphysics of Science? In S. Mumford, & M. Tugby (Eds.), *Metaphysics and Science* (pp. 3–26). Oxford: Oxford University Press.

Nerlich, G. (1994). *The Shape of Space* (2nd ed.). Cambridge: Cambridge University Press.

Newton, I. (1846). *The Mathematical Principles of Natural Philosophy*. (A. Motte, Trans.) New York, NY: Daniel Adee.

Paul, L. A. (2012). Metaphysics as modeling: The handmaiden's tale. *Philosophical Studies*, *160*, 1–29. doi:10.1007/s11098-012-9906-7

Rickles, D., & French, S. (2006). Quantum Gravity Meets Structuralism: Interweaving Relations in the Foundations of Physics. In D. Rickles, S. French, & J. Saatsi (Eds.), *The Structural Foundations of Quantum Gravity* (pp. 1–39). Oxford: Oxford University Press.

Rovelli, C. (2001). Quantum Spacetime: What Do We Know? In N. Huggett, & C. Callender (Eds.), *Physics Meets Philosophy at the Planck Scale. Contemporary Theories in Quantum Gravity* (pp. 101–122). Cambridge: Cambridge University Press.

Salimkhani, K. (2021). Explaining unification in physics internally. *Synthese, 198*, 5861–5882. doi:10.1007/s11229-019-02436-x

Salmon, W. C.. (1984). *Scientific Explanation and the Causal Structure of the World*. Princeton, NJ: Princeton University Press.

Salmon, W. C. (1990). Scientific explanation: Causation and unification. *Crítica: Revista Hispanoamericana de Filosofía, 22(66)*, 3–23.

Steinberger, F. (2017). The Normative Status of Logic. In E. N. Zalta (Ed.), *The Stanford Encyclopedia of Philosophy* (Spring 2017 ed.). Metaphysics Research Lab, Stanford University. https://plato.stanford.edu/archives/spr2017/entries/logic-normative

Treder, H.-J. (1983). "Was Gott getrennt hat, soll der Mensch nicht vereinen." Zum Problem der Großen Unitarisierung. *Astronomische Nachrichten, 304(4)*, 145–151. doi:10.1002/asna.2113040402

von Weizsäcker, C. F.. (1980). *The Unity of Nature*. New York, NY: Farrar, Straus and Giroux.

Weinberg, S. (1999). What Is Quantum Field Theory, and What Did We Think It Was? In T. Y. Cao (Ed.), *Conceptual Foundations of Quantum Field Theory* (pp. 241–251). Cambridge: Cambridge University Press.

Wüthrich, C. (2005). To quantize or not to quantize. Fact and folklore in quantum gravity. *Philosophy of Science, 72(5)*, 777–788. doi:10.1086/508946

7 Conclusion

In this work, I studied different aspects of the question whether spacetime is fundamental and concluded that on the basis of various approaches to quantum gravity (including apparently spatiotemporal ones) *and* general relativity, metrical aspects of spacetime are not fundamental, but derivative on matter field dynamics.

In Chapter 2, I first argued in general terms that an entity is fundamental if it is ontologically independent, and then discussed how this translates to the context of the philosophy of spacetime – for example, with respect to the traditional substantivalism–relationalism debate. Also, I pointed out that there are different aspects of spacetime that need to be distinguished; most importantly, metrical and topological aspects, the former of which I have mainly focused on here.

In Chapter 3, I then investigated in detail what special and general relativity can tell us about the fundamentality of spacetime. I argued that, according to the standard view, both theories suggest that all aspects of spacetime are fundamental. I then showed how Brown and Pooley's dynamical approach to relativity theory begins to challenge this conclusion. On a dynamical view, key spatiotemporal properties are taken as ontologically *dependent* on matter field dynamics, which calls into question spacetime fundamentalism. However, a further assessment of the dynamical approach reveals a couple of caveats, especially with respect to the dynamical approach to general relativity. The dynamical approach to general relativity demonstrates that chronogeometricity is an *extrinsic* property of the g field, but does not ontologically reduce the g field as such to matter field dynamics – everything else about g (e.g., its symmetry properties, its gravitational properties, and its having energy and momentum) is viewed as ontologically independent. As a result, the dynamical approach to general relativity has to accept *two* unexplained miracles. In addition (based on joint work with Niels Linnemann), I presented the problem of pregeometry, which points out that the status of the topological structure is opaque in the dynamical approach. As a solution, we propose that the

DOI: 10.4324/9781003404149-7

manifold is actually dispensable. Norton's central counter-argument only establishes the indispensability of point coincidences; point coincidences suffice if modelled in terms of interactions.

In Chapter 4, I then considered an alternative formulation of general relativity, namely classical spin-2 theory, and argued how spin-2 theory helps to resurrect the Brownian dynamical approach: the dynamically interpreted spin-2 theory explains the second miracle and renders metrical aspects non-fundamental due to the fact that, in spin-2 theory, the g field is ontologically reduced to matter fields (including a new field *h*) and their dynamics. I argued that the dynamically interpreted spin-2 theory is explanatorily more coherent than the standard geometrical reading of general relativity. Its ontological import, however, is less clear-cut. That being said, I do argue that we still have good reasons to view g as non-fundamental on the basis of classical spin-2 theory.

The fact that there is a reading of g as a non-fundamental entity at the level of classical general relativity foreshadows and reinforces the results in the context of (speculative) theories of quantum gravity, which I investigated in Chapter 5. Here I studied some of the reasons why the various research programmes of quantum gravity are even being considered, and discussed the general problems of specifying their potential metaphysical implications. I then turned to a more thorough investigation of one of the proposals in particular: the relativistic quantum particle approach to quantum gravity, i.e., the quantum version of spin-2 theory. Most importantly, I argued that the relativistic quantum particle approach yields a robust reformulation of the equivalence principle. Given the presumption of Lorentz invariance, however, this should not be considered a full-fledged derivation – contrary to frequent remarks in the physics literature. Furthermore, I highlighted that, while all results from the quantum spin-2 approach also obtain in its classical version, it is the quantum spin-2 approach which breaks what might have been considered an "ontological tie" between general relativity and classical spin-2 theory, and refutes spacetime fundamentalism: the fundamental ontology is significantly constrained by quantum spin-2 theory. In particular, metrical aspects of spacetime are unambiguously derivative on the dynamics of quantum matter including gravitons; the direction of the ontological dependence relation, namely "g ontologically depends on gravitons (and not *vice versa*)", is firmly fixed – unlike the respective dependence relation in classical spin-2 theory. Nevertheless, I argued that, ultimately, all theories and all tenable interpretations of theories which are compatible with the available empirical data are relevant for determining ontological inferences. Accordingly, the geometrical interpretations of classical general relativity and classical spin-2 gravity still push against the inference that metrical aspects of spacetime are non-fundamental, especially given the fact that the quantum spin-2 approach has its own problems (it is inconsistent at high energies) and that the other approaches to quantum gravity are not worked out in full.

In Chapter 6, I therefore attempted to resolve this issue by taking into account physics unificatory practise, which is naturally related to investigating fundamentality as independence. I first argued that unification is not an external constraint imposed on physics for metaphysical or metatheoretical reasons, but an internal *result* of everyday research in physics (brought about by the basic methodological strategies of physics). Based on these insights, I then proposed to constrain ontological inferences in the philosophy of spacetime by means of continuity conditions to, for example, fundamental particle physics; notably, the applicability of these continuity conditions to the context of spacetime philosophy needs to be established by concrete physical arguments, not some general appeal to simplicity or unification. Most importantly, it is quantum spin-2 theory which establishes an explicit theoretical connection between particle physics and general relativity by reducing general relativity to quantum field theory. As a result, interpretations of general relativity, which do not square well with the continuity conditions of particle physics (e.g., that there is a quantum spin-2 field that couples universally and that metrical aspects cannot be fundamental due to the Heisenberg collapse argument), should not be considered as having ontological import. In particular, spacetime fundamentalism cannot be considered a tenable ontological position. But interpretations of general relativity, which *do* fit the continuity conditions, that is the dynamical spin-2 interpretation of general relativity, do have ontological import and contribute to constraining to what extent spatiotemporality is non-fundamental – notably, pending further empirical results.

Let me conclude by briefly outlining a few remaining issues for future research. On the physics side, for example, the limitations of spin-2 theory and, more generally, its relation to general relativity need to be investigated more carefully. Also, the role of consistency requirements and possible additional tacit assumptions in the derivation of general relativity requires further examination. For instance, to what extent is general relativity unique as a non-linear spin-2 theory? Also, the similarities and differences between the classical and the quantum version of spin-2 theory need to be worked out in more detail. On the philosophy side, for example, one could explore how to reconstruct the various arguments presented in this book more explicitly as inferences to the best explanation. In addition, the different concepts of fundamentality and their metaphysical as well as methodological implications for the philosophy of physics deserve further attention. In this context, it would also be worthwhile to examine more closely the demarcation issue between debates on fundamentality and debates on reality.

Appendix
More Detailed Calculations

A.1 Reversing the Spin-2 Argument

As mentioned in Section 4.1.2, it is instructive to look at the reversed version of the classical spin-2 derivation of Einstein's field equations. The following presentation is taken from Weinberg (1972, pp. 165–171).

We can write Einstein's equations "in an entirely equivalent form that, because not manifestly covariant, reveals their relation to the wave equations of elementary particle physics" (Weinberg, 1972, p. 165). Take

$$g_{\mu\nu} = \eta_{\mu\nu} + h_{\mu\nu}, \tag{A.1}$$

with the classical spin-2 field $h_{\mu\nu}$ being assumed to vanish at infinity, but not being small everywhere (just as before in Chapter 4).

Consider the Ricci tensor,

$$R_{\mu\kappa} = g^{\lambda\nu} R_{\lambda\mu\nu\kappa}, \tag{A.2}$$

with the curvature tensor

$$R_{\lambda\mu\nu\kappa} = \frac{1}{2}\left[\frac{\partial^2 g_{\lambda\nu}}{\partial x^\kappa \partial x^\mu} - \frac{\partial^2 g_{\mu\nu}}{\partial x^\kappa \partial x^\lambda} - \frac{\partial^2 g_{\lambda\kappa}}{\partial x^\nu \partial x^\mu} + \frac{\partial^2 g_{\mu\kappa}}{\partial x^\nu \partial x^\lambda}\right] + g_{\eta\sigma}\left[\Gamma^\eta_{\nu\lambda}\Gamma^\sigma_{\mu\kappa} - \Gamma^\eta_{\kappa\lambda}\Gamma^\sigma_{\mu\nu}\right], \tag{A.3}$$

and the Christoffel symbols of the affine connection $\Gamma^\alpha_{\beta\gamma}$.

The part of $R_{\mu\kappa}$ that is linear in $h_{\mu\nu}$ computes to

$$R^{(1)}_{\mu\kappa} \equiv \frac{1}{2}\left[\frac{\partial^2 h^\lambda_{\ \lambda}}{\partial x^\mu \partial x^\kappa} - \frac{\partial^2 h^\lambda_{\ \mu}}{\partial x^\lambda \partial x^\kappa} - \frac{\partial^2 h^\lambda_{\ \kappa}}{\partial x^\lambda \partial x^\mu} + \frac{\partial^2 h_{\mu\kappa}}{\partial x^\lambda \partial x_\lambda}\right], \tag{A.4}$$

where indices are being raised and lowered by the Minkowski metric, for example, $h^\lambda_{\ \lambda} \equiv \eta^{\lambda\nu} h_{\lambda\nu}$ and $\partial/\partial x_\lambda \equiv \eta^{\lambda\nu} \partial/\partial x^\nu$. Of course, for $R_{\mu\nu}$, etc. indices are still raised and lowered with gs.

Equipped with the above definitions we can now rewrite the full Einstein field equations as follows:

$$R_{\mu\kappa}^{(1)} - \frac{1}{2}\eta_{\mu\kappa}R^{(1)\lambda}_{\ \lambda} = -8\pi G\left[T_{\mu\kappa} + t_{\mu\kappa}\right] \tag{A.5}$$

with

$$t_{\mu\kappa} \equiv \frac{1}{8\pi G}\left[R_{\mu\kappa} - \frac{1}{2}g_{\mu\kappa}R^{\lambda}_{\lambda} - R_{\mu\kappa}^{(1)} + \frac{1}{2}\eta_{\mu\kappa}R^{(1)\lambda}_{\ \lambda}\right] \tag{A.6}$$

Let me stress again that Eq. (A.5) is merely a rewriting of the standard Einstein field equations.

The form of Eq. (A.5) is that of a wave equation for a classical spin-2 field. Note, however, that the source term $T_{\mu\kappa} + t_{\mu\kappa}$ depends explicitly on the spin-2 field $h_{\mu\nu}$ itself. This can be interpreted as the spin-2 field being sourced by the *total* energy-momentum content – including a contribution of h itself, namely the energy-momentum "tensor" of the gravitational field $t_{\mu\nu}$. In particular, one can then define the total energy-momentum "tensor" of matter and gravitation as (Weinberg, 1972, p. 165):

$$\tau^{\nu\lambda} \equiv \eta^{\nu\mu}\eta^{\lambda\kappa}\left[T_{\mu\kappa} + t_{\mu\kappa}\right] \tag{A.7}$$

Weinberg supports this interpretation by demonstrating that various expected properties actually do hold (Weinberg, 1972, pp. 166–171). In particular, $\tau^{\mu\nu}$ is locally conserved in the ordinary sense, i.e.,

$$\frac{\partial}{\partial x^{\mu}}\tau^{\mu\nu} = 0 \tag{A.8}$$

whereas $T^{\mu\nu}$ "obeys the covariant conservation law $T^{\nu\lambda}_{\ ;\nu} = 0$, which really describes the *exchange* of energy between matter and gravitation" (Weinberg, 1972).

Weinberg concludes by pointing out that one could reverse this argumentation and "set out to construct equations for a long-range field of spin-2" to "provide yet another derivation of Einstein's field equations" (Weinberg, 1972, p. 171) – this is precisely what Chapter 4 is about.

A.2   Weinberg's Soft Graviton Argument

This is to provide a few more steps for Weinberg's calculation in Section 5.2.2. We start from the modified process of Figure 5.1b for the spin-1 case:

$$\mathcal{M}_{\alpha\beta}\left(p_1,\ldots,p_n,q\right)\big|_{q\to 0} = \mathcal{M}_{\alpha\beta}\left(p_1,\ldots,p_n\right)\times\sum_i e_i\,\frac{p_i^{\mu}}{2p_i\cdot q}\,\varepsilon_{\mu}\left(q\right) \tag{A.9}$$

and demand, as discussed, that $\mathcal{M}_{\alpha\beta}(p_1,\ldots,p_n,q)\big|_{q\to 0}$ is invariant under Lorentz transformations Λ, i.e.,

$$\mathcal{M}_{\alpha\beta}(p_1,\ldots,p_n,q)\big|_{q\to 0} \overset{!}{=} \mathcal{M}_{\alpha'\beta'}(p_1',\ldots,p_n',q')\big|_{q'\to 0} \tag{A.10}$$

$$= \Lambda^{\alpha}_{\alpha'}\Lambda^{\beta}_{\beta'}\mathcal{M}_{\alpha\beta}(\Lambda p_1,\ldots,\Lambda p_n,\Lambda q)\big|_{\Lambda q\to 0} \tag{A.11}$$

Since the non-modified amplitude $\mathcal{M}_{\alpha\beta}(p_1,\ldots,p_n)$ is already Lorentz-invariant, i.e.,

$$\mathcal{M}_{\alpha\beta}(p_1,\ldots,p_n) = \mathcal{M}_{\alpha'\beta'}(p_1',\ldots,p_n') \tag{A.12}$$

the photon emission factor

$$\mathcal{E}_{\text{Photon}} = \sum_i e_i \frac{p_i^{\mu}}{2p_i \cdot q}\varepsilon_{\mu}(q) \tag{A.13}$$

is constrained individually, i.e.,

$$\mathcal{E}_{\text{Photon}} \overset{!}{=} \mathcal{E}'_{\text{Photon}} = \sum_i e_i \frac{p_i'^{\mu}}{2p_i' \cdot q'}\varepsilon_{\mu}'(q) \tag{A.14}$$

Note that the polarisation vector $\varepsilon_{\mu}(q)$ is neither manifestly Lorentz-invariant, nor a four-vector (Weinberg, 1995, p. 537), but transforms into $(\Lambda\varepsilon)_{\mu}(q)$ *plus* a term proportional to the momentum q (Weinberg, 1995, p. 537):

$$\varepsilon_{\mu}(q) \to \varepsilon_{\mu}'(q) = (\Lambda\varepsilon)_{\mu}(q) + \alpha(\Lambda q)_{\mu} \tag{A.15}$$

i.e.,

$$\varepsilon_{\mu}(q) \to \varepsilon_{\mu}'(q) = \Lambda^{\sigma}_{\mu}\varepsilon_{\sigma}(q) + \alpha\Lambda^{\rho}_{\mu}q_{\rho} \tag{A.16}$$

So to obtain Lorentz invariance, we need to explicitly demand that

$$\varepsilon_{\mu}(q) \overset{!}{=} \varepsilon_{\mu}'(q) \tag{A.17}$$

$$= (\Lambda\varepsilon)_{\mu}(q) + \alpha(\Lambda q)_{\mu} \tag{A.18}$$

i.e., we need to introduce a *redundancy*. Note that Eq. (A.18) follows from two manifestly Lorentz-invariant constraints: (1) $p^{\mu}p_{\mu} = 0$ (the photon is massless) and (2) $\varepsilon_{\mu}(p)p^{\mu} = 0$ (the polarisation vector is, per definition, orthogonal to the photon's momentum) (Arkani-Hamed, 2010).

We can shift $\varepsilon_\mu(q)$ by anything proportional to q without changing the physical state, i.e., the amplitude (A.9) should give the same result for $\varepsilon_\mu(q)$ and $\varepsilon_\mu'(q)$; in other words, these states are in the same equivalence class: $\varepsilon_\mu(q) \sim \varepsilon_\mu(q) + \alpha q_\mu$ (Arkani-Hamed, 2010). Hence, Eq. (A.9) is required to become zero for replacing $\varepsilon_\mu(q)$ by q_μ. Since $\mathcal{M}_{\alpha\beta}(p_1, \ldots, p_n)$ is assumed to be non-zero (the original process is not forbidden), the emission factor itself has to vanish (as discussed in Section 5.2.2), and we arrive at the fact that the sum over all charges needs to be zero, $\Sigma_i e_i = 0$.

The graviton case is completely analogous, but uses the polarisation *tensor* $\varepsilon_{\mu\nu}(q)$. In particular, the polarisation tensor $\varepsilon_{\mu\nu}(q)$ is neither manifestly Lorentz-invariant, nor a true tensor, but transforms into $(\Lambda\Lambda\varepsilon)_{\mu\nu}(q)$ *plus* a term proportional to the corresponding momentum q times some generic vector α^ν (Arkani-Hamed, 2010), namely:

$$\varepsilon_{\mu\nu}(q) \to \varepsilon_{\mu\nu}'(q) = (\Lambda\Lambda\varepsilon)_{\mu\nu}(q) + (\Lambda q)_\mu (\Lambda\alpha)_\nu \tag{A.19}$$

So this again introduces a redundancy. We can shift $\varepsilon_{\mu\nu}(q)$ by q_μ multiplied by some generic vector α_ν without changing the physical state, i.e., the amplitude should give the same result for $\varepsilon_{\mu\nu}(q)$ and $\varepsilon_{\mu\nu}'(q)$; in other words, $\varepsilon_{\mu\nu}(q)$ and $\varepsilon_{\mu\nu}(q) + q_\mu \alpha_\nu$ are in the same equivalence class.

It is instructive to look at this from the perspective of degrees of freedom (again following Arkani-Hamed, 2010). As pointed out in Chapter 4, the whole argument is based on the fact that massless (i.e., $p^2 = p^\mu p_\mu = 0$) particles of any spin higher than spin-0 have only *two physically significant* spin degrees of freedom, dubbed helicities, which is at odds with the fact that they are, at first, *represented by more than two* degrees of freedom: in the spin-1 case, we have four-vectors with four degrees of freedom, and, in the spin-2 case, we have (symmetric) tensors with ten degrees of freedom. The reason that scalar theories are less restrictive than higher-spin theories – for example with respect to which energy-momentum tensors are possible (Ortín, 2015) – is that for scalars the representational degrees of freedom *do* match the physical degrees of freedom from the start.

Consider the spin-1 case. With respect to spin, a spin-1 particle is essentially represented by the polarisation four-vector $\varepsilon_\mu(p)$. Note that we can explicitly write down any four-vector in terms of a plane wave state $A_\mu(x) = \varepsilon_\mu(p)e^{ipx}$. Standardly, one spurious degree of freedom can be eliminated by noting that the polarisation vector is orthogonal to the particle's momentum vector, i.e., $\varepsilon_\mu(p)p^\mu = 0$. But it seems that we are stuck with the remaining three representational degrees of freedom, because, as we have seen, $\varepsilon_\mu(p)$ is not manifestly Lorentz-invariant: it transforms according to Eq. (A.15). However, by introducing the redundancy of Eq. (A.17) and thereby *explicitly demanding* Lorentz invariance, we can eliminate another degree of freedom and obtain that only two degrees of freedom are physically significant – as desired.

Now, consider the spin-2 case. Here we start from ten representational degrees of freedom for the symmetric tensor $h_{\mu\nu}$ (see Chapter 4). These are reduced to five independent degrees of freedom by noting that $h_{\mu\nu}$ is traceless, i.e.,

$$h^\mu_\mu = 0 \tag{A.20}$$

which gives one constraint, and by noting that

$$p^\mu h_{\mu\nu} = 0 \tag{A.21}$$

which gives four additional constraints. But we are still left with five degrees of freedom instead of two, such that three spurious degrees of freedom still need to be eliminated. We are again forced to introduce a redundancy by considering how $h_{\mu\nu}$ transforms under Lorentz transformations, namely

$$h_{\mu\nu} \rightarrow h'_{\mu\nu} = h_{\mu\nu} + \alpha_\mu p_\nu + \alpha_\nu p_\mu \tag{A.22}$$

and demand that $h_{\mu\nu}$ and $h'_{\mu\nu}$ represent the same physics. Note that, due to the trace condition for $h_{\mu\nu}$, the α_μ are not completely arbitrary, but need to satisfy $\alpha^\mu p_\mu = 0$. Alternatively, we could lift this constraint and forget about Eq. (A.20). So we get rid of three redundancy degrees of freedom, such that we are left with two physically relevant degrees of freedom, as desired. Notice that Eq. (A.22) is nothing but linearised diffeomorphism invariance (Arkani-Hamed, 2010). We could, in fact, continue and construct the linearised Riemann tensor (Arkani-Hamed, 2010).

References

Arkani-Hamed, N. (2010). Robustness of GR. Attempts to Modify Gravity. Part I. Prospects in Theoretical Physics Program: Frontiers of Physics in Cosmology, Institute for Advanced Study, 25–28 July 2011. https://video.ias.edu/pitp-2011-arkani-hamed1

Ortín, T. (2015). *Gravity and Strings*. Cambridge: Cambridge University Press.

Weinberg, S. (1972). *Gravitation and Cosmology: Principles and Applications of the General Theory of Relativity*. New York, NY: Wiley.

Weinberg, S. (1995). *The Quantum Theory of Fields. Volume I: Foundations.* Cambridge: Cambridge University Press.

Index

For Product Safety Concerns and Information please contact our EU
representative GPSR@taylorandfrancis.com
Taylor & Francis Verlag GmbH, Kaufingerstraße 24, 80331 München, Germany

* 9 7 8 1 0 3 2 5 1 8 3 4 3 *